W9-BYG-881

InspireScience

Be a Scientist Notebook

Student Journal

Grade 1

Mc
Graw
Hill
Education

mheducation.com/prek-12

Copyright © 2017 McGraw-Hill Education

All rights reserved. No part of this publication may be
reproduced or distributed in any form or by any means,
or stored in a database or retrieval system, without
the prior written consent of McGraw-Hill Education,
including, but not limited to, network storage or
transmission, or broadcast for distance learning.

STEM McGraw-Hill is committed to providing
instructional materials in Science, Technology,
Engineering, and Mathematics (STEM) that give all
students a solid foundation, one that prepares them
for college and careers in the 21st century.

Send all inquiries to:
McGraw-Hill Education
STEM Learning Solutions Center
8787 Orion Place
Columbus, OH 43240

ISBN: 978-0-07-678221-5
MHID: 0-07-678221-2

Printed in the United States of America.

1 2 3 4 5 6 7 8 9 QVS 21 20 19 18 17 16

Our mission is to provide educational resources that enable
students to become the problem solvers of the 21st century
and inspire them to explore careers within Science, Technology,
Engineering, and Mathematics (STEM) related fields.

TABLE OF CONTENTS

Sound Energy

Module Opener 2

Lesson 1 Sound 4

Lesson 2 Making Sounds 18

Module Wrap-Up 32

Light Energy

Module Opener 34

Lesson 1 Light and Shadows 36

Lesson 2 Properties of Light 48

Lesson 3 How Light Travels 62

Module Wrap-Up 72

Use Energy to Communicate

Module Opener 74

Lesson 1 Communicate With Light and Sound 76

Lesson 2 Communication Technology 88

Module Wrap-Up 102

Plants and Animals

Module Opener 104

Lesson 1 Living and Nonliving Things 106

Lesson 2 Parts of Plants 116

Lesson 3 Parts of Animals 126

Lesson 4 Plant and Animal Survival 140

Module Wrap-Up 152

DEVEN
Sound Engineer

TABLE OF CONTENTS

Offspring and Their Parents

Module Opener		154
Lesson 1	Plants Grow and Change	156
Lesson 2	Plants and Their Parents	170
Lesson 3	Compare Animals	182
Lesson 4	Animals and Their Parents	194
Lesson 5	Offspring and Survival	208
Module Wrap-Up		222

Earth and Space

Module Opener		224
Lesson 1	Day and Night	226
Lesson 2	Seasonal Patterns	242
Lesson 3	The Moon	256
Lesson 4	The Sun and Stars	270
Module Wrap-Up		284

VKV Visual Kinesthetic Vocabulary

Introduction	286
Visual Kinesthetic Vocabulary Cut-Outs	289

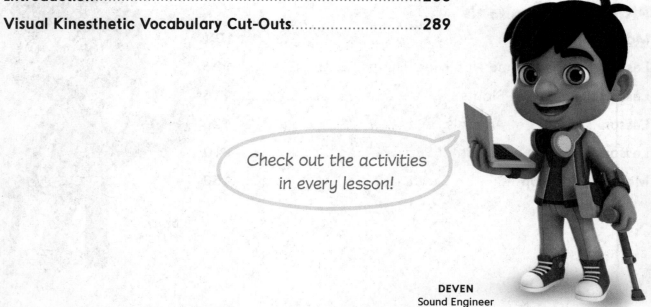

Check out the activities in every lesson!

DEVEN
Sound Engineer

Inspire Science

This is your own journal. You will be a scientist or an engineer. Write in your book as you answer questions and solve problems.

Draw a picture to show what a scientist or an engineer might do.

EMILY
Aerospace Engineer

Sound Energy

🌍 Science in My World

▶ Watch the video of the drums. Drums can make many different sounds. What questions do you have about drums and sound?

_____

- -

_____

_____

- -

_____

_____

- -

_____

abc Key Vocabulary

Look and listen for these words as you learn about sound energy.

energy	matter	pitch
sound	vibrate	volume
waves		

How do drums make sound?

ERIK
Video Game Designer

Erik wants to be a video game designer. Video game designers use sound to tell players what is happening in the game. Erik is curious about how he can use the drums to make sound effects for a video game. Drums can make loud sounds and soft sounds. Show how you think the drums make sound.

Science and Engineering Practices

I will plan an investigation.

I will carry out an investigation.

Sound

 **PAGE KEELEY SCIENCE PROBES**

Sound

Circle the things that make sound by vibrating.

Playing a guitar	**Two blocks rubbed together**
Blowing a horn	**Ringing a bell**
Beating a drum	**Person humming**
Air coming out of a balloon	**Bouncing ball**

Explain your thinking.

_____

- -

_____

- -

Science in My World

Look at the photo of the different instruments. What do you observe about how each instrument makes sound? What questions do you have?

_____

- -

_____

- -

_____

- -

_____

- -

_____

❓ Essential Question
How is sound made?

⚙️ Science and Engineering Practices

I will carry out an investigation.

> I love to record things that make noise. I wonder how each instrument makes different sounds!

DEVEN
Sound Engineer

Inquiry Activity
Rubber Band Guitar

How do vibrations make sound?

Make a Prediction How can you make sound with a rubber band?

_____

- -

_____

_____

- -

_____

_____

- -

_____

Materials

- [ ] safety goggles
- [ ] rubber bands
- [ ] bowl

Carry Out an Investigation

BE CAREFUL Wear goggles to protect your eyes.

1. Stretch a rubber band around the opening of the bowl.

2. Pluck the rubber band and listen to its sound.

You can make a guitar with a bowl and a rubber band!

Communicate Information

1. Record Data Use the table.

What I Saw	What I Heard

2. **Communicate** What made the sound in your activity?

_____

- -

_____

_____

- -

_____

3. Predict what your guitar would sound like with a thicker rubber band. Circle the answer.

a. The sound will be louder.

b. The sound will be lower.

c. The sound will be higher.

4. **Test** Follow the same procedure.
Use a thicker rubber band. How did
your predictions compare to your
observations?

_____
- -
_____
_____
- -
_____
_____
- -
_____
_____
- -
_____

5. **Draw Conclusions** What do you
think caused the rubber band to
make sound?

_____
- -
_____
- -
_____
- -
_____
- -
_____
- -
_____

Obtain and Communicate Information

Vocabulary

Use these words when explaining sound.

sound   energy   vibrate

matter   waves

Sound, Energy, and Matter

Read *Sound, Energy, and Matter* on how those three things are related. Answer the question after you have finished reading.

1. Match the word with its definition.

   sound   what all things are made of

   matter   form of energy that comes from objects that vibrate

Sound Waves

Watch *Sound Waves* on what makes a sound. Answer the questions after you have finished watching.

2. Fill in the blank.

_____

- - - - - - - - - - - - - - - -

A _____ makes matter move by vibrations.

3. (Circle) the statement that is true. Place an ✕ over the statement that is false.

You can see vibration but not sound waves.

You can see sound waves but not vibration.

Sounds All Around

📖 Read pages 14–17 in *Sounds All Around.* Answer the questions.

4. (Circle) the picture that shows where sound is made when you speak.

5. How does the cricket on page 17 make sounds?

_____

- -

_____

_____

- -

⚙ Science and Engineering Practices

Complete the "I can . . ." statement.

_____

I can carry out an investigation

_____

- -

_____

- -

_____

- -

Use examples from the lesson to explain what you can do!

Research, Investigate, and Communicate

Inquiry Activity
Sound Journal

You will listen to different sounds for one hour and suggest what causes each sound.

Ask a Question What do you want to learn about in your investigation?

_____

- -

_____

Carry Out an Investigation

Record Data Keep a list of the sounds you hear for one hour. What caused each sound?

Sound I Heard	Cause of the Sound

Communicate Information

1. **Communicate** Draw a picture showing one of the sounds you heard. Label the parts that caused the sound.

⚙ Crosscutting Concepts
Cause and Effect

2. What causes a rubber band to make sound?

_____

- -

_____

- -

_____

- -

_____

- -

⚙ Performance Task
Design an Instrument

You will use common objects to make your own instruments.

Make a Prediction What instruments can you make with common objects?

_____

- -

_____

_____

- -

_____

Materials

☐ safety goggles

☐ metal can

☐ cup

☐ bowl

☐ seeds/rice

☐ sticks

☐ spoons

☐ plastic containers with lids

Carry Out an Investigation

BE CAREFUL Wear goggles to protect your eyes.

① Look at some common objects. What materials will you use for your instrument? Check the box next to the materials you choose to use for your instrument.

② Think about how you can use the objects to make sound. Design an instrument using the objects.

3 **Test** How will you test your instrument?

_____

- -

_____

- -

_____

4 Draw your instrument. Label the parts that make sound.

Communicate Information

1. **Communicate** Describe the sound your object makes.

_____

- -

_____

- -

_____

2. Tell how your instrument makes sound.

_____

- -

_____

_____

- -

_____

3. **Draw Conclusions** Compare your instrument with others in your class. How does each instrument make sound?

Instrument	How Instrument Makes Sound

? Essential Question
How is sound made?

Think about the photo of people playing instruments at the beginning of the lesson. Use what you have learned to tell how sound is made.

_____

- -

_____

- -

_____

- -

_____

⚙ Science and Engineering Practices

I did carry out an investigation.

Rate Yourself

Color in the number of stars that tell how well you did carry out an investigation.

Now that you're done with the lesson, rate how well you did.

Making Sounds

PAGE KEELEY
SCIENCE
PROBES

Materials and Vibrations

Which friend has the best idea about sound?

Jeffrey

Hector

Lona

Lona: I think sound can make materials vibrate.

Jeffrey: I think materials can make sound when they vibrate.

Hector: I think sound can make materials vibrate and vibrating materials can make sound.

Explain your thinking.

_____

- -

_____

- -

_____

🌎 Science in My World

Look at the photo of the elephants. Listen to the sounds they make. What questions do you have about how the elephants make sounds?

_____

- -

_____

- -

_____

- -

_____

- -

_____

❓ Essential Question
How does sound change?

Elephants can make many different kinds of sounds. I wonder how sounds can change!

⚙️ Science and Engineering Practices

I will plan an investigation.

Inquiry Activity
Sound Waves

How can sound waves change?

Make a Prediction How are sounds different?

_____

- -

_____

_____

- -

_____

Materials

☐ safety goggles

☐ plastic cups or bowls

☐ water

☐ tuning forks

Carry Out an Investigation

BE CAREFUL Wear goggles to protect your eyes.

1 Pour some water into your cup or bowl.

2 Strike the tuning fork gently and listen to its sound.

3 Place the tuning fork in the water and look at the ripples.

What do you see? What do you hear?

Communicate Information

1. **Record Data** Use the table. Write what you saw and what you heard.

What I Saw	What I Heard

2. **Communicate** What made the sound in your activity?

_____

- -

_____

- -

_____

- -

_____

3. **Test** Follow the same procedure, but strike the tuning fork harder. Tell what happened.

_____

- - - - - - - - - - - - - - - - - -

_____

- - - - - - - - - - - - - - - - - -

_____

- - - - - - - - - - - - - - - - - -

_____

- - - - - - - - - - - - - - - - - -

_____

4. **Draw Conclusions** How are the sounds and the waves related?

_____

- - - - - - - - - - - - - - - - - -

_____

- - - - - - - - - - - - - - - - - -

_____

- - - - - - - - - - - - - - - - - -

_____

- - - - - - - - - - - - - - - - - -

_____

🗨️ Obtain and Communicate Information

🔤 Vocabulary

Use these words when explaining sound.

volume   pitch

Instruments

🎛️ Investigate different sounds by conducting the simulation. Answer the question after you have finished.

1. Fill in the blank.
   _____

   The more an object vibrates, the _____ the sound.

Different Sounds

👁️ Read *Different Sounds* on volume and pitch. Answer the question after you have finished reading.

2. Match the word with its definition.

   pitch   how loud or soft a sound is

   volume   how high or low a sound is

Sounds All Around

📖 Read pages 18–23 in *Sounds All Around*.

3. Which animal would make sounds with the lowest pitch? (Circle) the answer.

lion   mouse   cricket

4. Deven hears a soft, high-pitched sound. What do you know about the vibrations?

_____

- -

_____

- -

_____

Use examples from the lesson to explain what you can do!

⚙️ Science and Engineering Practices

Complete the "I can . . ." statement.

I can plan an investigation

- -

_____

- -

_____

🔍 Research, Investigate, and Communicate

✋ Inquiry Activity
Throat Vibrations

You will observe the vibrations of your voice.

Make a Prediction How will the vibrations of your throat change when you make different sounds with your voice?

_____

- -

_____

- -

_____

Carry Out an Investigation

1. Think of different ways you can use your voice to make sound.

2. Place your hand on your throat. Describe what you feel when you use your voice in different ways.

I can make lots of different sounds using my voice! How do the vibrations change for each sound?

Communicate Information

1. **Record Data** Describe what you feel in each of the situations below. Use the table.

What You Are Doing	What You Are Feeling
Not talking	
Talking	
Whispering	
Shouting	
Laughing	

2. Draw Conclusions Tell how the vibrations of your throat would change if you made sounds in a lower pitch.

_____

- - - - - - - - - - - - - - - - - - - -

_____

- - - - - - - - - - - - - - - - - - - -

_____

- - - - - - - - - - - - - - - - - - - -

_____

- - - - - - - - - - - - - - - - - - - -

_____

Crosscutting Concepts
Cause and Effect

3. How does sound make matter move?

_____

- - - - - - - - - - - - - - - - - - - -

_____

- - - - - - - - - - - - - - - - - - - -

_____

- - - - - - - - - - - - - - - - - - - -

_____

- - - - - - - - - - - - - - - - - - - -

_____

⚙ Performance Task
Sound Energy

You will use water and tuning forks to compare the vibrations of low sounds and high sounds.

Ask a Question What do you want to learn about pitch and vibrations?

_____

- - - - - - - - - - - - - - - - - - -

_____

_____

- - - - - - - - - - - - - - - - - - -

_____

_____

- - - - - - - - - - - - - - - - - - -

_____

Carry Out an Investigation
What materials will you use?

_____

- - - - - - - - - - - - - - - - - - -

_____

_____

- - - - - - - - - - - - - - - - - - -

_____

_____

- - - - - - - - - - - - - - - - - - -

_____

What steps will you use to do your investigation?

① _____

② _____

③ _____

Communicate Information

1. Record Data Use the table.

What I Saw	What I Heard

2. Draw Conclusions Tell what you learned about pitch and vibrations in your investigation.

_____

- -

_____

- -

_____

- -

_____

- -

_____

Circle the word True or False to describe the statement.

3. A sound with a low volume has a strong vibration.

True   False

4. A sound with a high pitch vibrates quickly.

True   False

? Essential Question
How does sound change?

Think about the photo of elephants at the beginning of the lesson and the different sounds the elephants make. Use what you have learned to tell how sound changes.

_____

_____

_____

_____

_____

_____

⚙ Science and Engineering Practices

I did plan an investigation.

Rate Yourself

Color in the number of stars that tell how well you did plan an investigation.

Now that you're done with the lesson, rate how well you did.

Name _____ Date _____

Sound Energy

⚙ Performance Project
Sound and Matter

Materials

- [ ] safety goggles
- [ ] plastic bowls
- [ ] plastic wrap
- [ ] salt
- [ ] music speaker

You will investigate how sound affects salt on a plastic-wrapped bowl.

Make a Prediction How can sound make salt move?

_____

- -

_____

- -

_____

Carry Out an Investigation

What steps will you use?

_____

- -

① _____

_____

- -

② _____

_____

- -

③ _____

To make sound, matter needs to vibrate.

Communicate Information

Record Data Write or draw your observations.

[blank box for observations]

🌎 Explore More in My World

Did you learn the answers to all of your questions from the beginning of the module? If not, how could you design an experiment or conduct research to help answer them?

Light Energy

🌍 Science in My World

Look at the photo of light shining through
the windows. What do you observe about
the light and the windows? What questions
do you have about light?

_____

- -

_____

- -

_____

- -

_____

abc Key Vocabulary

Look and listen for these words as you
learn about light energy.

light	materials	mirror
opaque	reflect	shadow
translucent	transparent	

How does light show an object?

MALIK
Photonics Engineer

Malik wants to be a photonics engineer. Photonics engineers develop and design technologies that use light energy. Malik is curious about the colored windows. How does light travel through the windows but not the window frames? Show how you think light travels.

⚙ Science and Engineering Practices

I will construct an explanation.
I will design a solution.

Light and Shadows

PAGE KEELEY SCIENCE PROBES

Light and Sight
Which friend has the best idea about light?

Finn

Jane

Finn: *I think we need light to see things.*

Jane: *I think we can see some things without light.*

Explain your thinking.

_____

- -

_____

- -

_____

Science in My World

Look at the photo of the shadow of a tree. Is light needed to make that shadow? What do you want to know about the shadow?

_____

- -

_____

- -

_____

- -

_____

- -

_____

? Essential Question
What is light?

Science and Engineering Practices

I will construct an explanation.
I will design a solution.

I am planning the set for the school play. I need to think about light to make sure the actors and their props can be seen!

CHLOE
Carpenter

Inquiry Activity
Shadow Walk

Materials
☐ paper
☐ pencil

What causes shadows?

Make a Prediction How are shadows different?

_____

- -

_____

- -

_____

- -

_____

Carry Out an Investigation

BE CAREFUL Follow your teacher's instructions.

1. Take a walk outside the building on a sunny day.

2. Look for your shadow and observe its shape.

3. Look for shadows made by other objects.

4. Observe the shapes of those shadows.

Observe which objects make shadows.

Communicate Information

1. Record Data Make a list of
the shadows you saw.

_____

_____

_____

_____

2. Communicate What happened to
your own shadow when you moved?

_____

_____

_____

_____

3. Draw a picture of one of the shadows you saw. Label the objects making the shadows. Label the Sun.

⚙ Crosscutting Concepts
Cause and Effect

4. What caused the shadows?

_____

- -

_____

_____

- -

_____

_____

- -

💬 Obtain and Communicate Information

🔤 Vocabulary

Use these words when explaining light.

light   shadow

Lights and Shadows

📖 Read pages 14–23 in *Lights and Shadows.*
Answer the questions after you have
finished reading.

1. ⬭Circle the pictures that show sources
of light. Place an ✕ over the pictures
that are not sources of light.

 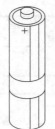 

2. How does light help us?

_____

- -

_____

How Light and Shadows Interact

Explore the Digital Interactive
How Light and Shadows Interact
on light and shadows.

3. Fill in the blank.

When an object blocks light,
_____

- - - - - - - - - - - - - - -
a _____ happens.

Use examples
from the lesson to
explain what you
can do!

Science and Engineering Practices

Complete the "I can . . ." statements.

I can construct an explanation

I can design a solution

Research, Investigate, and Communicate

Inquiry Activity
Changing Shadows

You will observe the shapes of shadows during the day.

Make a Prediction How does a shadow change during the day?

_____

- -

_____

- -

_____

- -

_____

- -

_____

Carry Out an Investigation

① Choose an object that makes a shadow outside.

② Pick three different times of day to observe the shadow.

What does your shadow look like at different times of the day?

Communicate Information

1. **Record Data** Draw a picture of the shadow made by your object at three times during the day. Use the table.

Time of Day	What the Shadow Looks Like

2. **Draw Conclusions** Why did the shadow change during the day?

_____

- -

_____

- -

_____

- -

_____

Performance Task
Lighting the School Play

You will make a model and write instructions for lighting the school play.

Define a Problem You are helping Chloe to plan the sets and lighting for the school play. What do you need to consider?

Materials

- [ ] shoe box
- [ ] construction paper
- [ ] scissors
- [ ] craft sticks
- [ ] tape or glue
- [ ] flashlights

_____

- -

_____

- -

_____

- -

_____

Make a Model

1 **Make a model of the stage and the sets. Include the scenery, props, and actors.**

2 **Test** Test your model in a dark room. Use flashlights as the spotlights.

Help me plan for the school play. What do I need to consider about the sets and lighting?

Communicate Information

1. Tell about your model. Can you see all
of the actors and props?

_____
_____
_____
_____
_____
_____

2. **Communicate** What do you need
to consider about the scenery and
props you are using? Where should
the spotlights be? Write a set
of instructions to use for creating
the sets and lighting the show.

_____
_____
_____
_____
_____
_____
_____
_____

❓ Essential Question
What is light?

Think about the photo of a tree's shadow at the beginning of the lesson. Use what you have learned to tell what light is.

_____

- -

_____

- -

_____

- -

_____

- -

_____

⚙️ Science and Engineering Practices

I did construct an explanation.
I did design a solution.

Rate Yourself

Color in the number of stars that tell how well you did construct an explanation and design a solution.

> Now that you're done with the lesson, rate how well you did.

Properties of Light

PAGE KEELEY SCIENCE PROBES

Properties of Light
Which friend has the best idea about light?

Paolo   Kit   Lily

Paolo: Light can pass through all materials.

Kit: Light can pass through some materials.

Lily: Light does not pass through any materials.

Explain your thinking.

_____

_____

_____

_____

Science in My World

Look at the photo of light shining through the leaf. What do you observe about the light? What questions do you have?

_____

- -

_____

- -

_____

- -

_____

- -

_____

- -

_____

? Essential Question

How does light travel through different materials?

Science and Engineering Practices

I will plan an investigation.

I want to learn why light shines through some materials and not others.

✋ Inquiry Activity
Light Passing Through

Which items allow light to pass through?

Make a Prediction Which items will let light pass through them?

_____

- -

_____

- -

_____

- -

_____

Materials
☐ safety goggles
☐ cardboard tube
☐ aluminum foil
☐ wax paper
☐ plastic wrap
☐ rubber bands

Carry Out an Investigation

BE CAREFUL Handle the materials carefully.

1. Look through the cardboard tube.

2. Cover the end of the tube with aluminum foil. Use a rubber band to keep the foil in place. Look through the tube.

3. Repeat step 2 with wax paper and then again with plastic wrap.

Communicate Information

1. Record Data Use the table to tell if you can see light through each item.

Item	Observations
None	
Aluminum Foil	
Wax Paper	
Plastic Wrap	

2. Communicate Which items let light pass through?

_____

- -

_____

- -

_____

3. How could you learn which other items will let light pass through?

_____

- -

_____

_____

- -

_____

4. Draw Conclusions Why do you think light can pass through some items and not others?

_____

- -

_____

- -

_____

- -

_____

- -

_____

💬 Obtain and Communicate Information

🔤 Vocabulary

> Use these words when explaining properties of light.
>
> materials   transparent   opaque
>
> translucent

Light

▶ Watch *Light* on how light shines through different materials. Answer the following questions after you have finished watching.

1. Match the word with the object it describes.

opaque	clear window
transparent	orange slice
translucent	wood board

2. Fill in the blanks.

If you can see through material,

_____

- - - - - - - - - - - - - - - - - - -

it is _____.

To keep light out completely,

_____

- - - - - - - - - - - - - - - - - - -

use _____ material.

Types of Materials

Explore the Digital Interactive *Types of Materials* on light and different materials.

3. Translucent materials allow all light to pass through them. (Circle) the answer.

True   False

4. A brick wall is an example of an opaque material. (Circle) the answer.

True   False

5. Tell about a time when you used a transparent, translucent, or opaque material? Why?

_____

- -

_____

⚙ Crosscutting Concepts
Cause and Effect

6. What causes light to pass through some materials and not others?

_____

- -

_____

- -

_____

- -

_____

⚙ Science and Engineering Practices

Complete the "I can . . ." statement.

_____

- -

I can plan an investigation

- -

_____

- -

_____

- -

_____

Use examples from the lesson to explain what you can do!

Research, Investigate, and Communicate

Inquiry Activity
Building Materials

You will observe different materials in your school that are transparent, translucent, and opaque.

> Look at the different objects around your classroom or school. Why are different materials used for different objects?

Carry Out an Investigation

1. Go for a walk around your classroom or your school building. Look for objects and parts of the building that use materials that are transparent, translucent, and opaque.

2. Observe the objects and materials, and think about how and why each is used.

Communicate Information

1. Draw one example of each type of material that you saw.

Transparent	Translucent	Opaque

2. List an example of each type of material you saw. Tell how the material is used. Use the table.

Type of Material	Example	How Material Is Used
Transparent		
Translucent		
Opaque		

3. Draw Conclusions Why is it helpful that different parts of your school or objects in your classroom are built with different materials?

_____

- -

_____

- -

_____

⚙ Performance Task
Light and Materials

You will plan an investigation to find out if light passes through different materials.

What materials do you want to test in your investigation?

_____

_____

_____

_____

_____

Make a Prediction Which materials will let light pass through? Which will block light?

_____

_____

_____

_____

_____

_____

Materials
☐ safety goggles
☐ cardboard tube
☐ rubber bands
☐ materials to test

Carry Out an Investigation

What steps should you use to do your investigation?

_____

- -

_____

- -

_____

- -

_____

Communicate Information

1. Record Data Use the table.

Material	Observations

2. **Draw Conclusions** How are the objects that block light and the objects that let light pass through the same? How are they different?

_____

- -

_____

_____

- -

_____

_____

- -

_____

_____

- -

_____

3. Chloe wants to make sure her room is dark so she can sleep late tomorrow morning. List the materials she should use or not use.

_____

- -

_____

_____

- -

_____

_____

- -

_____

_____

- -

_____

? Essential Question
How does light travel through different materials?

Think about the photo of light shining through the leaf at the beginning of the lesson. Use what you have learned to tell how light travels through different materials.

_____

- -

_____

_____

- -

_____

Science and Engineering Practices

I did plan an investigation.

Rate Yourself

Color in the number of stars that tell how well you did plan an investigation.

Now that you're done with the lesson, rate how well you did.

How Light Travels

 **PAGE KEELEY SCIENCE PROBES**

Mirrors and Light

What will happen when the boy shines the flashlight on the mirror?

☐ The light will go through the mirror.

☐ The light will bounce off the mirror.

☐ The mirror will stop the light.

Explain your thinking.

_____

- -

_____

- -

_____

Science in My World

Look at the photo of the mirrored ball. What happens when light hits the mirrored ball? What questions do you have?

_____

_____

_____

_____

_____

? Essential Question
How can light bounce off objects?

⚙ Science and Engineering Practices

I will carry out an investigation.

I want to learn about how light travels and how light can bounce off objects.

Find the Cat

🔡 Investigate what happens to light when objects are in its path by conducting the simulation. Answer the questions after you have finished.

1. What happened when the light was shined directly on the cat? Write or draw your answer.

2. What happened when the light traveled to a mirror? Write or draw your answer.

3. Which objects allowed light to pass through? Which objects did not allow light to pass through?

_____

- - - - - - - - - - - - - - - - - - - -

_____

- - - - - - - - - - - - - - - - - - - -

_____

- - - - - - - - - - - - - - - - - - - -

_____

- - - - - - - - - - - - - - - - - - - -

_____

4. What causes some objects to allow light to pass through?

_____

- - - - - - - - - - - - - - - - - - - -

_____

- - - - - - - - - - - - - - - - - - - -

_____

- - - - - - - - - - - - - - - - - - - -

_____

- - - - - - - - - - - - - - - - - - - -

_____

💬 Obtain and Communicate Information

🔤 Vocabulary

> Use these words when explaining how light travels.
>
> mirror   reflect

How Does Light Move?

👁 Read *How Does Light Move?* on how light travels.

1. List two sources of light.

_____

- -

_____

2. How does light travel? (Circle) the answer.

    a. Light travels in a straight line.

    b. Light travels in circles.

    c. Light travels back and forth.

Mirrors and Light

📖 Read pages 14–23 in *Mirrors and Light.* Answer the questions after you have finished reading.

3. Fill in the blank.

_____

- - - - - - - - - - - - - - - - -

A _____ reflects or gives back an image of an object.

⚙ Crosscutting Concepts
Cause and Effect

4. How does light travel with a mirror in its path? Draw your answer.

⚙ Science and Engineering Practices

Complete the "I can . . ." statement.

_____

I can carry out an

investigation _____

- -

Use examples from the lesson to explain what you can do!

Research, Investigate, and Communicate

A Prism

▶ Watch *A Prism* to see how light changes when it passes through a prism. Answer the questions after you have finished.

1. Circle the statement that is true. Place an ✕ over the statement that is false.

   A prism can make light visible.

   A prism can split visible light into colors.

2. Draw what happens when white light passes through a prism.

Performance Task

Mirrors

You will plan an investigation to find out if light passes through a mirror.

Make a Prediction What happens when light hits a mirror?

_____

- -

_____

Carry Out an Investigation

What materials will you use?

_____

- -

_____

- -

_____

What steps will you use to do your investigation?

_____

- -

_____

- -

_____

- -

_____

Communicate Information

1. Communicate What did you observe?
Draw or write your answer.

2. Draw Conclusions What conclusions
can you make about light and mirrors?

_____

- -

_____

_____

- -

_____

- -

_____

- -

_____

? Essential Question
How can light bounce off objects?

Think about the photo of the mirrored ball at the beginning of the lesson. Use what you have learned to tell how light bounces off objects.

_____

_____

_____

_____

_____

⚙ Science and Engineering Practices

I did carry out an investigation.

Now that you're done with the lesson, rate how well you did.

Rate Yourself

Color in the number of stars that tell how well you did carry out an investigation.

Light Energy

⚙️ Performance Project
Light Illuminates Objects

You will investigate how light can show an object.

Make a Prediction How can you see an object inside a pinhole box?

_____

- -

_____

- -

_____

Materials
☐ shoe box or other small box
☐ scissors
☐ pin or nail
☐ flashlight
☐ small objects

Carry Out an Investigation

1. Build a pinhole box. Make a small hole on a short side of the box. Make a hole to shine the flashlight through on one of the long sides of the box.

2. Place an object in the box.

3. Shine the flashlight through the hole on the long side of the box. Look through the hole on the short side of the box.

We can see objects only when light shines on them.

Communicate Information

1. **Record Data** What do you see? Draw or write your observations.

2. Is the object illuminated on all sides?

_____

- -

_____

- -

_____

 Explore More in My World

Did you learn the answers to all of your questions from the beginning of the module? If not, how could you design an experiment or conduct research to help answer them?

Use Energy to Communicate

🌍 Science in My World

▶ Watch the video of the ambulance.
It uses lights and sirens to communicate.
What questions do you have about
the ambulance?

_____

- -

_____

_____

- -

_____

_____

- -

_____

🔤 Key Vocabulary

Look and listen for these words as you
learn about using energy to communicate.

communicate   electricity   receiver

transmitter

How do you communicate?

MARISOL
Paramedic

Marisol wants to be a paramedic. A paramedic is a person who helps in an emergency. Marisol is curious about the ambulance. How do emergency vehicles use light and sound to communicate? Draw or write your answer.

Science and Engineering Practices

I will construct an explanation.
I will design a solution.

Communicate with Light and Sound

PAGE KEELEY
SCIENCE
PROBES

Tin Can Message Sender

Two children tied a long string to two tin cans. They used this device to send and get messages. How do you think they sent and got messages with this device?

☐ They used light.

☐ They used sound.

☐ They used electricity.

Explain your thinking.

_____

- -

_____

_____

- -

_____

Science in My World

Look at the photo of the lighthouse. What do you think the light is communicating? What questions do you have about the lighthouse?

_____

- -

_____

_____

- -

_____

- -

_____

- -

_____

? Essential Question
How do we use light and sound to communicate?

I want to learn about light and sound. What different ways do people use light and sound to communicate?

⚙ Science and Engineering Practices

I will construct an explanation.

ERIK
Video Game Designer

✋ Inquiry Activity
Communicate with Light and Sound

Which objects use light to communicate?
Which objects use sound to communicate?

Materials

☐ objects or pictures of objects that use light and/or sound

Make a Prediction How do some objects communicate with people?

_____

- -

_____

- -

_____

- -

_____

Carry Out an Investigation

① Look at each object.

② Think about how the object communicates with people. Does it use light, sound, or both?

③ Sort the objects into groups to tell if they use light, sound, or both to communicate.

How do objects use light and sound to communicate?

Communicate Information

1. Record Data Use the table.

Light	Sound	Both Light and Sound

2. Communicate What is an object you often use to communicate using light and sound?

_____

- -

_____

- -

_____

- -

_____

- -

_____

Obtain and Communicate Information

Vocabulary

Use these words when explaining communication with light and sound.

communicate   electricity

Lighthouses

▶ Watch *Lighthouses* on how a lighthouse communicates with light and sound.

1. How does the lighthouse help people? (Circle) the answer.

    a. It tells people on land that water is near.

    b. It shows the people on the ships that land is near.

    c. It lights up the beach so people can go swimming at night.

Light and Sound Are Energy

📖 Read pages 14–23 in *Light and Sound Are Energy*. Answer the following questions.

2. Circle the pictures of items that use electricity. Place an ✕ on items that do not use electricity.

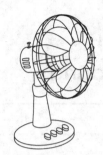

⚙ Crosscutting Concepts
Influence of Engineering, Technology, and Science on Society and the Natural World

3. How do you use light and sound to communicate every day? How would your life be different without using light and sound to communicate?

_____

_____

_____

_____

_____

_____

_____

_____

Explore the Digital Interactive *Use Sound and Light to Communicate* on ways we use sound and light every day.

4. Why does a smoke detector flash a bright light and make a loud sound?

_____

- -

_____

_____

- -

_____

Science and Engineering Practices

Complete the "I can . . ." statement.

_____

I can construct an explanation

- -

_____

- -

_____

Use examples from the lesson to explain what you can do!

- -

_____

Research, Investigate, and Communicate

Inquiry Activity
Games with Light and Sound

Materials

☐ video games or board games that use light and sound

You will investigate how games use light and sound.

Make a Prediction How do games use light and sound?

_____

- -

_____

_____

- -

_____

Design a Solution

Think of a video game or board game that uses light and sound. How does the game use light and sound?

_____

- -

_____

_____

- -

_____

You can design your own game that uses light and sound. How would you use light and sound in your game?

Communicate Information

1. Design your own game. How would
   it use sound, light, or both to
   communicate? Tell about your game.

_____

- -

_____

_____

- -

_____

_____

- -

_____

2. Draw what your game would look like.
   Label the parts that use light or sound.

⚙ Performance Task
Paper Cup Phone

You will investigate how the length of string affects sound traveling between the paper cups.

Make a Prediction Which string will make the best sound? Why?

_____

- -

_____

_____

- -

_____

Materials
☐ meterstick
☐ 1 m string
☐ 3 m string
☐ 7 m string
☐ pencil
☐ paper cups
☐ paper clips

Carry Out an Investigation

1. Measure each string with the meterstick. Put the strings in order from shortest to longest.

2. Push one string through the holes of the paper cups so one cup is on each end of the string.

3. Tie the ends of the string around paper clips to keep the string in place.

4. Use your phone to talk and listen to each other. Repeat steps 3–5 for each string.

Which string will make the best sound for the paper cup phone?

Communicate Information

1. Record Data What was the length of the shortest string you used? What was the length of the longest string you used?

_____

- -

_____

- -

_____

2. Draw Conclusions How did the strings affect the phone and communication?

_____

- -

_____

- -

_____

3. Design a Solution How could you make your paper cup phone better? Draw a model of your improved phone. Label the materials you would use.

❓ Essential Question

How do we use light and sound to communicate?

Think about the photo of the lighthouse at the beginning of the lesson. Use what you have learned to tell how people communicate using light and sound.

_____

- -

_____

- -

_____

⚙️ Science and Engineering Practices

I did construct an explanation.

Rate Yourself

Color in the number of stars that tell how well you did construct an explanation.

Now that you're done with the lesson, rate how well you did.

Communication Technology

 PAGE KEELEY
SCIENCE
PROBES

Rescue at Sea

A man in a lifeboat sees a ship far away.
He has a flashlight and a horn in the lifeboat.
What do you think is the best way for him to
send a message to the ship?

☐ He should use the flashlight.

☐ He should blow the horn.

Explain your thinking.

_____

- -

_____

- -

_____

Science in My World

Look at the photo of the astronaut. How do you think astronauts communicate or talk with people on Earth? What questions do you have about how the astronaut communicates?

_____

- -

_____

- -

_____

- -

_____

? Essential Question

How has communication changed over time to make people's lives easier?

> I am curious about the astronaut. How do astronauts communicate with people on Earth?

⚙ Science and Engineering Practices

I will design a solution.

✋ Inquiry Activity
Making Megaphones

How does a megaphone affect sound?

Make a Prediction What does a megaphone do to sound?

_____

- -

_____

_____

- -

_____

_____

- -

_____

Materials
☐ fabric
☐ foil
☐ poster board
☐ printer paper
☐ tape or glue

Carry Out an Investigation

BE CAREFUL Handle the materials carefully. Follow your teacher's instructions.

1. Choose the material you think will make the best megaphone. Draw your design on the next page.

2. Shape the material into a megaphone and secure it with tape or glue.

3. **Test** Use the megaphone to focus sound across the room.

Communicate Information

1. Draw a model of your megaphone.
Label the materials. Label the length
of your megaphone. Label the size of
the openings of your megaphone.

2. Communicate How did your
megaphone compare to ones made
by your classmates?

_____

- -

_____

3. How could you make your
megaphone better?

_____

- -

_____

Obtain and Communicate Information

🔤 Vocabulary

Use these words when explaining communication technology.

transmitter   receiver

Communication Over Time

▶ Watch *Communication Over Time* on ways communication has changed.

1. Tell one way people communicated with each other before phones and computers were invented.

_____

- -

_____

- -

_____

- -

_____

- -

_____

The Design Process

👁 Read *The Design Process* on how scientists design solutions to problems.

2. Place a 1, 2, 3, or 4 in the box to put the steps of the design process in order.

☐ Draw, plan, and build a solution.

☐ Share the design with others.

☐ Think of a problem.

☐ Test the solution.

Alexander Graham Bell and the Telephone

👁 Read *Alexander Graham Bell and the Telephone* on how the telephone was invented. Answer the questions after you have finished reading.

3. What part of the telephone carried the electric current with the sound?

a. the wires

b. the transmitter

c. the receiver

4. How did Alexander Graham Bell use the design process?

_____

- - - - - - - - - - - - - - - - - - -

_____

- - - - - - - - - - - - - - - - - - -

_____

_____

- - - - - - - - - - - - - - - - - - -

_____

Technology Helps People

📖 Explore the Digital Interactive *Technology Helps People* on different ways people use technology to communicate.

5. Which is the best way to alert a visually impaired person that it is safe to cross the street? Circle the answer.

a. Use sound signals.

b. Use flashing lights.

c. Use pictures.

⚙ Crosscutting Concepts
Influence of Engineering, Technology, and Science on Society and the Natural World

6. It is difficult for some people to speak or see. How does technology help them communicate with others?

_____

- -

_____

- -

_____

- -

_____

- -

_____

- -

_____

- -

_____

- -

_____

Science and Engineering Practices

Complete the "I can . . ." statement.

I can design a solution

Use examples from the lesson to explain what you can do!

🔍 Research, Investigate, and Communicate

Megaphones

🗔 Investigate *Megaphones* by conducting the simulation. Answer the questions after you have finished.

1. What shape and length made the best megaphone?

_____
- -
_____
_____
- -
_____

2. Who might use a megaphone to communicate at their job? What problem could a megaphone solve?

_____
- -
_____
_____
- -
_____
_____
- -

 Performance Task
Send Messages

Materials

☐ flashlight with batteries

You will research and investigate how Morse Code helps people communicate information over a long distance.

Define a Problem What problem will you try to solve using Morse Code?

_____

- -

_____

- -

_____

_____

- -

_____

- -

Carry Out an Investigation

① **Research** Find out about Morse code and how it was used.

② **Design** a way to send Morse Code messages using light.

③ **Test** Send a message to a classmate using your solution. Receive a message from a classmate.

> How can you use Morse Code and light to send messages?

Communicate Information

1. Communicate What did you learn about Morse Code? What is it? How was it used on ships?

_____

- -

_____

- -

_____

- -

_____

A ●—	J ●———	S ●●●
B —●●●	K —●—	T —
C —●—●	L ●—●●	U ●●—
D —●●	M ——	V ●●●—
E ●	N —●	W ●——
F ●●—●	O ———	X —●●—
G ——●	P ●——●	Y —●——
H ●●●●	Q ——●—	Z ——●●
I ●●	R ●—●	

2. Write some words using Morse Code.

_____

- -

Cat _____

- -

Dog _____

- -

Hello _____

3. Tell what materials you used in your solution.

_____
- -
_____
_____
- -
_____

4. Describe how you used Morse Code and light to send messages.

_____
- -
_____
_____
- -
_____

5. Draw Conclusions How can people still use Morse Code with light today? What problem could you use Morse Code to solve?

_____
- -
_____
_____
- -
_____
_____
- -
_____

❓ Essential Question

How has communication changed over time to make people's lives easier?

Think about the photo of the astronaut at the beginning of the lesson. Use what you have learned to tell how communication technology helps make our lives easier.

_____
- -
_____
- -
_____
- -
_____

⚙️ Science and Engineering Practices

I did design a solution.

Rate Yourself

Color in the number of stars that tell how well you did design a solution.

Now that you're done with the lesson, rate how well you did.

Use Energy to Communicate

⚙️ Performance Project
Design a Communication Device

You will design a device that uses light, sound, or both to communicate.

Define a Problem How can you communicate using light or sound to solve a problem at your school? You may want to interview classmates to learn about problems at your school that could be solved by using light or sound energy.

_____

- -

_____

Design a Solution How can a device that uses light, sound, or both solve the problem? Tell about your solution.

_____

- -

_____

There are many ways to communicate information with light and sound energy!

Draw a picture of your device. Label the parts that use light or sound. Tell what materials you need to create your device. Tell how you could test your device.

 Explore More in My World

Did you learn the answers to all of your questions from the beginning of the module? If not, how could you design an experiment or conduct research to help answer them?

Plants and Animals

Science in My World

Look at the photo of the airplane. Humans
use airplanes to travel. Airplanes can fly.
Animal and plant parts can be a model for
machines that humans use. What questions
do you have about the airplane?

_____

- -

_____

_____

- -

abc Key Vocabulary

Look and listen for these words as you
learn about plants and animals.

flower	fruit	leaves
living	mammal	nonliving
nutrient	root	seed
shelter	stem	survive

How did nature influence the design of the airplane?

EMILY
Aerospace Engineer

Emily wants to be an aerospace engineer. An aerospace engineer is a person who designs airplanes and spaceships. Show how you think looking at plants and animals influenced the design of airplanes. Draw an example below.

Science and Engineering Practices

I will construct an explanation.
I will design a solution.

Living and Nonliving Things

PAGE KEELEY
SCIENCE
PROBES

Living and Nonliving

Circle the things that are living.

Tree	**Fire**	**Seeds**
Cloud	**Rabbit**	**Waterfall**
Icicle	**Beetle**	**Bird**

Explain your thinking.

_____

- -

_____

- -

_____

Science in My World

▶ Watch the video of the jellyfish in the ocean. Are the jellyfish living things? What questions do you have about jellyfish?

_____

- -

_____

- -

_____

- -

_____

? Essential Question
How are living and nonliving things different?

I love to learn about things that live in the ocean! I wonder how I can tell what things are living and what things are nonliving.

⚙ Science and Engineering Practices

I will construct an explanation.

HIRO
Ocean Engineer

🖐 Inquiry Activity
What Seeds Need to Grow

What do seeds need in order to grow into a plant?

Make a Prediction What do seeds need to grow?

_____
- -
_____
_____
- -
_____

Materials
☐ two small cups
☐ soil
☐ small rock
☐ package of seeds
☐ black marker
☐ water

Carry Out an Investigation

BE CAREFUL Handle the materials carefully.

① Fill each cup halfway with soil.

② Place the rock in one cup and a seed in the other cup. Label each cup.

③ Fill each cup with more soil.

④ Place both cups in a sunny location.

⑤ Add water to each cup once a week.

⑥ Observe the cups once a week for four weeks.

Communicate Information

1. Record Data Use the table.

	Week 1	Week 2	Week 3	Week 4
Rock				
Seeds				

2. Communicate What did the seed
need to grow?

_____

_____

_____

_____

----

3. Draw Conclusions Which cup holds
a living thing? Tell why it is living.

_____

_____

_____

_____

_____

🗨 Obtain and Communicate Information

🔤 Vocabulary

Use these words when explaining living and nonliving things.

living   nonliving

Living and Nonliving Things

🔄 Explore the Digital Interactive *Living and Nonliving Things* on how living and nonliving things are different.

1. Fill in the blank.

   Things that need food, water, and air
   _____
   to grow and change are _____.

   Things that do not grow and change
   _____
   are _____.

2. Trees are not living things because they do not move.

   True   False

What Is Living and Nonliving?

Explore the Digital Interactive *What Is Living and Nonliving?*

3. Draw a living thing and a nonliving thing.
   Tell why each is living or nonliving.

Living	Nonliving

⚙️ Science and Engineering Practices

Complete the "I can . . ." statement.

Use examples from the lesson to explain what you can do!

I can construct an

explanation

_____

_____

Research, Investigate, and Communicate

What Living Things Need

Research You will research a living thing to learn what it needs to survive.

Ask a Question What question will your research help to answer?

_____

- -

_____

Communicate Information

1. What does your living thing need to live? Write a "how-to" list on a separate piece of paper to tell how to care for your living thing.

Crosscutting Concepts
Structure and Function

2. Which body parts of your living thing help it get what it needs to live?

_____

- -

_____

- -

_____

Glue your "how-to" list here.

⚙ Performance Task
Tell What is Living and Nonliving

Look at the pictures. Circle the things that are living. Put an ✕ on the things that are nonliving.

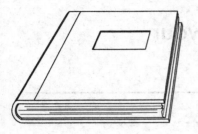

❓ Essential Question
How are living and nonliving things different?

▶ Think about the video of the jellyfish at the beginning of the lesson. Use what you have learned to tell how living and nonliving things are different.

_____

- -

_____

- -

_____

- -

_____

⚙️ Science and Engineering Practices

I did construct an explanation.

Now that you're done with the lesson, rate how well you did.

Rate Yourself

Color in the number of stars that tell how well you did construct an explanation.

Parts of Plants

PAGE KEELEY
SCIENCE
PROBES

Plant Parts

Circle the parts of a plant that help it live and grow.

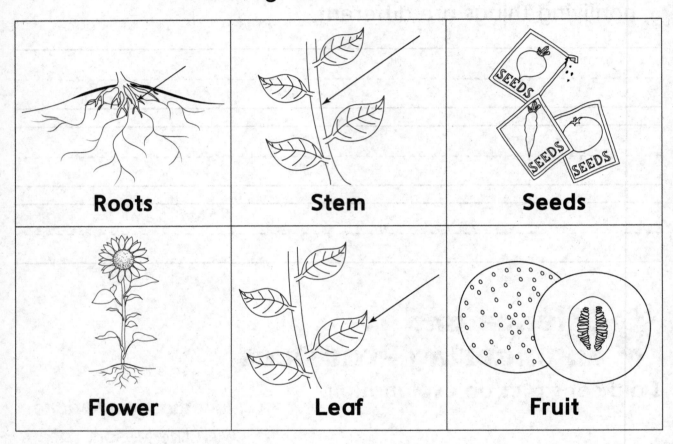

Roots	**Stem**	**Seeds**
Flower	**Leaf**	**Fruit**

Explain your thinking.

_____

- -

_____

- -

_____

Science in My World

Look at the photo of the lobster claw plant. What do you observe about the parts of the plant? What questions do you have?

_____

- -

_____

- -

_____

- -

_____

- -

_____

? Essential Question
How do different parts of a plant help it live?

⚙ Science and Engineering Practices

I will construct an explanation.

I love to learn about plants. I wonder how different parts of a plant help it stay alive!

KAYLA
Landscape Architect

✋ Inquiry Activity
Parts of a Plant

What are the parts of plants?

Make a Prediction Do all plants have the same parts?

_____

- -

_____

- -

_____

- -

_____

Materials
☐ safety goggles
☑ hand lens
☐ daisy
☐ onion

Carry Out an Investigation

BE CAREFUL Wear safety goggles to protect your eyes.

1. Use the hand lens to examine the root, stem, and leaves of each plant.

2. Compare the parts of the two plants.

Observe the parts of each plant. Do both plants have the same parts?

Communicate Information

1. Record Data What plant parts did you observe? Draw them in the table.

	Root	**Stem**	**Leaf**
Daisy			
Onion			

2. How are the plants' parts alike and different?

_____

- -

_____

Obtain and Communicate Information

abc Vocabulary

Use these words when explaining parts of plants.

flower	seed	stem
leaves	roots	nutrients
fruit		

The Parts of a Plant

 Explore the Digital Interactive *The Parts of a Plant* on different plant parts.

1. Match each plant part with its function.

   flower   holds up the plant

   stem   use sunlight and air to make food

   leaves   makes seeds or fruit

   roots   take in water and nutrients from soil

Plants are Living Things

📖 Read *Plants Are Living Things* pages 14–23.

2. Choose a plant part. Tell what it does for the plant.

_____

The _____ helps the

_____

plant _____ .

3. What part of the plant grows around seeds? Circle the answer.

a. leaves

b. roots

c. fruit

Use examples from the lesson to explain what you can do!

Science and Engineering Practices

Complete the "I can . . ." statement.

I can construct an explanation

Research, Investigate, and Communicate

Plants in Different Environments

Research You will research a plant in an environment that is different from where you live.

Ask a Question What question will your research help to answer?

_____

- -

_____

Record Data Research a plant that lives in the environment. Draw a picture of the plant and its environment.

Communicate Information

1. How does the plant live in the environment?

_____

- - - - - - - - - - - - - - - - - - - -

_____

_____

- - - - - - - - - - - - - - - - - - - -

_____

2. Could this plant live in your neighborhood? Explain.

_____

- - - - - - - - - - - - - - - - - - - -

_____

_____

- - - - - - - - - - - - - - - - - - - -

_____

Crosscutting Concepts
Structure and Function

3. How do plant parts help plants live and grow?

_____

- - - - - - - - - - - - - - - - - - - -

_____

_____

- - - - - - - - - - - - - - - - - - - -

_____

Performance Task
Plant Model

You will design a model of a plant.

Materials
- [ ] craft sticks
- [ ] construction paper
- [ ] scissors
- [ ] glue

Make a Model

Design a model of a plant. Include the different parts that help the plant survive.

Communicate Information

What parts does your plant have? Explain how each part helps the plant survive in its environment.

_____

_____

_____

_____

_____

_____

_____

_____

_____

_____

? Essential Question
How do different parts of a plant help it live?

Think about the photo of the lobster claw plant at the beginning of the lesson. Use what you have learned to tell how plant parts help plants live.

_____

_____

_____

_____

_____

Science and Engineering Practices

I did construct an explanation.

Rate Yourself

Color in the number of stars that tell how well you did construct an explanation.

Now that you're done with the lesson, rate how well you did.

Parts of Animals

**PAGE KEELEY
SCIENCE
PROBES**

Do They Have Body Parts?

Circle the animals that have body parts to help them live and grow.

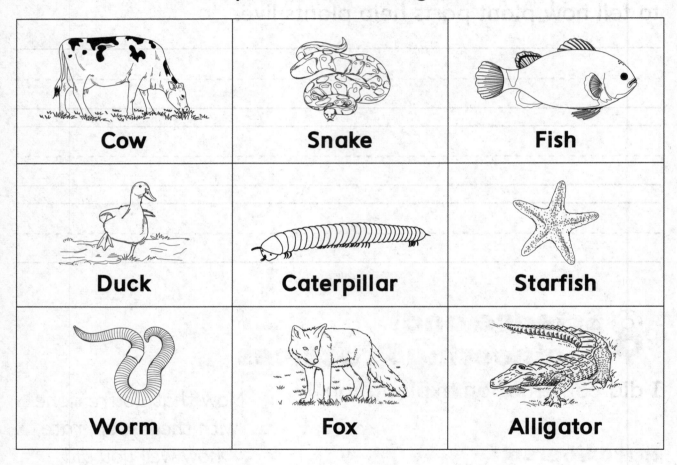

Cow	**Snake**	**Fish**
Duck	**Caterpillar**	**Starfish**
Worm	**Fox**	**Alligator**

Explain your thinking.

_____

- -

_____

- -

_____

Science in My World

▶ Watch the video to see a sea turtle swimming in a tropical coral reef. What body parts help the turtle to swim? What questions do you have?

_____

- -

_____

- -

_____

_____

- -

_____

? Essential Question
How do body parts help animals?

> I want to learn more about animals. I wonder how the different body parts of the sea turtle help it get what it needs.

⚙ Science and Engineering Practices

I **will** construct an explanation.

🖐 Inquiry Activity
What Animal Am I?

How can you investigate the different characteristics of animals?

Make a Prediction What body parts of animals help them to stay alive?

_____

- -

_____

- -

_____

Carry Out an Investigation

① **Form a large circle with your classmates.**

② **Classmates will take turns stepping into the circle so everyone can see the picture on their back.**

③ **When a classmate is in the circle, describe the parts of the animal pictured on his or her back.**

④ **Repeat the procedure with another classmate until all students have had a turn in the center of the circle.**

Describe the body parts of the animal in the picture.

Communicate Information

1. What body parts do animals have? Draw a picture of your animal. Think about what it looks like, how it moves, and where it lives.

2. Draw a picture of a classmate's animal. Think about what it looks like, how it moves, and where it lives.

3. Complete the Venn diagram to compare and contrast the two animals you drew.

Animal 1   Animal 2

Obtain and Communicate Information

abc Vocabulary

Use these words when explaining animals.

mammal	bird	reptile
amphibian	fish	insect
gills	lungs	

Animals Are Living Things

📖 Read pages 14–23 in *Animals Are Living Things.* Answer the questions after you have finished reading.

1. What type of animal lives part of its life in water and part on land? (Circle) the answer.

   a. a fish

   b. an insect

   c. an amphibian

2. Which animal is a mammal? Circle the picture. Tell how you know.

_____

- -

_____

_____

- -

_____

⚙ Crosscutting Concepts
Structure and Function

3. How do animals use their body parts to get what they need? Give an example.

_____

- -

_____

_____

- -

_____

_____

- -

_____

Animal Parts

Explore the Digital Interactive *Animal Parts* on how different body parts help a polar bear live. Answer the question after you have finished.

4. Fill in the blanks.

_____

- - - - - - - - - - - - - - - - -

A polar bear has _____ to see and find food. It uses its

_____

- - - - - - - - - - - - - - - -

_____ to catch and eat food.

_____

- - - - - - - - - - - - - - -

A polar bear has _____ to keep it warm in cold temperatures.

_____

- - - - - - - - - - - - - - - -

A polar bear uses its _____ to hear and sense danger.

⚙️ Science and Engineering Practices

Complete the "I can . . ." statement.

I can construct an explanation

Use examples from the lesson to explain what you can do!

🔍 Research, Investigate, and Communicate

Materials

☐ photos of different animals

☐ graph paper

✋ Inquiry Activity
Sorting Animals

You will sort photos of animals into groups that have similar parts.

Ask a Question What do you want to learn about animal parts?

_____

- -

_____

_____

- -

_____

Carry Out an Investigation

1 Look at photos of different animals.

2 Sort the photos into groups of animals that have similar parts.

> What different ways can you group the animals?

Communicate Information

1. Record Data Use the graph paper to make a bar graph of your results.

2. How many total animals did you have?

_____

- -

_____

3. Analyze Data Compare your animal groups. Which group had the most? How many more animals were in the biggest group than were in the smallest group?

_____

- -

_____

- -

_____

- -

_____

- -

_____

- -

_____

- -

_____

Glue your graph here.

⚙ Performance Task
Animal Parts

You will show how an animal uses its body parts to get what it needs.

Make a Prediction Which body parts help an animal get what it needs?

_____

- -

_____

- -

_____

- -

_____

Make a Model
1. Choose an animal. Research what your animal needs to live. Where does it live? What does it eat? Tell about your animal.

_____

- -

_____

- -

_____

- -

_____

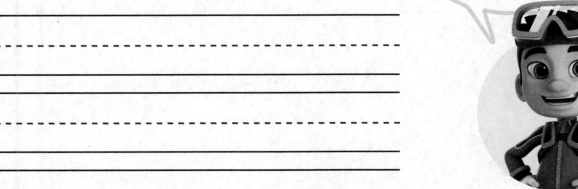

How does your favorite animal use its body parts to get what it needs?

2. Draw your animal. Label the parts
that help it get what it needs to live.

Communicate Information

3. List three body parts your animal has.
Tell how it uses each part to get what
it needs. Use the table.

Animal Part	How It Helps the Animal

? Essential Question
How do body parts help animals?

▶ Think about the video of the sea turtle at the beginning of the lesson. Use what you have learned to tell how body parts help animals survive.

_____

- -

_____

_____

- -

_____

_____

- -

_____

⚙ Science and Engineering Practices

I did construct an explanation.

Now that you're done with the lesson, rate how well you did.

Rate Yourself
Color in the number of stars that tell how well you did construct an explanation.

Plant and Animal Survival

PAGE KEELEY
SCIENCE
PROBES

Sensing Things

Which friend has the best idea about senses?

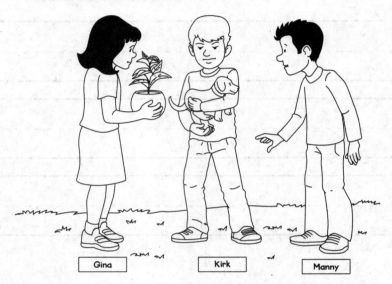

Gina   Kirk   Manny

Gina: I think only plants sense things around them.

Kirk: I think only animals sense things around them.

Manny: I think both plants and animals sense things around them.

Explain your thinking.

_____

- -

_____

- -

_____

🌍 Science in My World

▶ Watch the video of the Venus flytrap. What is the plant doing? What questions do you have?

_____

- -

_____

- -

_____

- -

_____

- -

_____

- -

_____

❓ Essential Question

How can plant and animal parts help us solve human problems?

> I like to learn about parts of different plants. How can plants give us ideas for solving human problems?

⚙️ Science and Engineering Practices

I will design a solution.

KAYLA
Landscape Architecht

🖐 Inquiry Activity
Bugs and Light

How do ground beetles respond to light?

Make a Prediction What will ground beetles do when a light shines on them?

_____

- -

_____

- -

_____

- -

_____

Carry Out an Investigation

📖 **Investigate how beetles respond to light by conducting the simulation. Answer the questions after you have finished.**

① Turn the lights on and off.

② Observe how the beetles respond.

Observe how the beetles respond when you turn the lights on and off.

Communicate Information

1. Record Data How many bugs were in the light area? How many bugs were in the dark area? Graph your results.

Number of Bugs

Light Area   Dark Area

2. Analyze Data What do your results show? How does the behavior of the bugs help them survive?

_____

- -

_____

- -

_____

- -

_____

- -

_____

- -

_____

💬 Obtain and Communicate Information

🔤 Vocabulary

Use these words when explaining plant and animal survival.

survive   shelter

Animals are Living Things

📖 Read pages 20–24 in *Animals Are Living Things*.

1. Fill in the blank.

   A place where animals can live and be

   _____

   - - - - - - - - - - - - - - - -

   safe is called a _____.

Ways Animals Use Their Senses

👁 Read *Ways Animals Use Their Senses*.

2. How do animals use their senses
   to survive?

   _____

   -

   _____

Plant Structure and Function

⚙ Explore the Digital Interactive *Plant Structure and Function* on how plant parts are similar to building parts. Answer the questions after you have finished.

3. What plant part is similar to the foundation of a building which holds the building in place? (Circle) the correct answer.

a. stem

b. leaves

c. roots

⚙ Crosscutting Concepts
Structure and Function

4. How is a plant stem similar to part of a building?

_____

_____

_____

Animal Structure and Function

Explore the Digital Interactive *Animal Structure and Function* on how some tools people use are similar to structures found in nature that help plants or animals survive.

5. Tell about one tool people use that is similar to an animal part. How does it help people? How does it help the animal?

_____
- -
_____
_____
- -

Use examples from the lesson to explain what you can do!

Science and Engineering Practices

Complete the "I can . . ." statement.

I can design a solution

🔍 Research, Investigate, and Communicate

✋ Inquiry Activity
Solving Human Problems

You will solve a human problem by mimicking a plant or animal part that helps the plant or animal survive.

Define a Problem Tell what problem you want to solve.

_____

- -

_____

_____

- -

_____

Design a Solution

Draw your solution to the problem.
Tell how it solves the problem.

Communicate Information

1. Draw a picture of the plant or animal part that helped you think of a solution.

2. Tell how your solution is like the plant or animal part.

_____

- -

_____

- -

_____

- -

_____

- -

 Performance Task
Design a New Tool

You will construct a model of a device that is similar to a Venus flytrap and that can solve a human problem.

Define a Problem What problem would you like your device to solve?

_____

_____

_____

Design a Solution

1. Make a model of your device. Tell how your device would work.

_____

_____

_____

_____

_____

2. Draw your device. Label its parts.

Communicate Information

3. Describe how your device mimics
the Venus flytrap.

_____

- -

_____

_____

- -

_____

_____

- -

_____

_____

- -

_____

? Essential Question

How can plant and animal parts help us solve human problems?

▶ Think about the video of the Venus flytrap at the beginning of the lesson. Use what you have learned to tell how plant and animal parts help us solve human problems.

_____

- -

_____

- -

Science and Engineering Practices

I did design a solution.

Rate Yourself

Color in the number of stars that tell how well you did design a solution.

Now that you're done with the lesson, rate how well you did.

Plants and Animals

⚙️ Performance Project
Nature-Inspired Tools

You will design a solution to a human problem by mimicking how plants or animals use their external parts to help them survive, grow, and meet their needs.

Design a Solution

Melanie and her family are going on a camping trip. They need to pack food, clothing, and a tent. Design something that her family can use to carry these things to their campsite. Draw your design below.

Things in nature are the inspiration for many human designs!

What thing in nature inspired your design?

_____
- -
_____
- -
_____
_____
- -
_____
_____
- -
_____
_____
- -
_____

 ## Explore More in My World

Did you learn the answers to all of your questions from the beginning of the module? If not, how could you design an experiment or conduct research to help answer them?

Offspring and Their Parents

🌍 Science in My World

Look at the photo of the tadpole. Tadpoles are young frogs. How will they grow and change? What questions do you have about tadpoles?

_____
- -
_____
_____
- -
_____
_____
- -
_____

🔤 Key Vocabulary

Look and listen for these words as you learn about offspring and their parents.

adaptation	behavior	carnivore
characteristic	herbivore	inherit
learn	life cycle	seedling
signal	trait	young

How do tadpoles grow and change?

POPPY
Park Ranger

Poppy wants to be a park ranger. Park rangers look after the plants and animals that live in public parks and make sure that they are not disturbed by people who visit the parks. Poppy wants to know more about how tadpoles grow and change. Show how you think the tadpoles will grow and change. Draw a picture of it below.

⚙ Science and Engineering Practices

I will obtain information.
I will communicate information.

Plants Grow and Change

PAGE KEELEY
SCIENCE
PROBES

Growing Plants

Which friend has the best idea about plants?

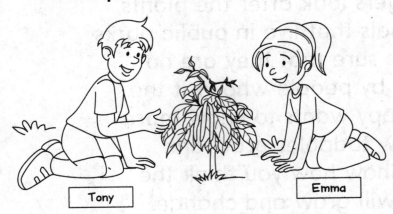

Tony

Emma

Emma: I think plants change in many ways as they grow.

Tony: I think plants don't change very much as they grow. They just look bigger.

Explain your thinking.

_____

- -

_____

_____

- -

_____

_____

- -

_____

Science in My World

▶ Watch the video of a plant starting to grow from a seed. How is the seed changing? What questions do you have about the plant?

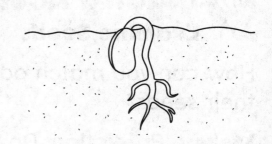

_____

- -

_____

- -

_____

- -

_____

- -

_____

- -

? Essential Question

How do plants grow and change?

⚙ Science and Engineering Practices

I will construct an explanation.

I love gardening with all different kinds of plants! I like to see how plants grow and change!

KAYLA
Landscape Architect

✋ Inquiry Activity
Examine Seeds

How can you match adult plants with their seeds?

Make a Prediction Do all plants grow from the same kinds of seeds?

_____

- -

_____

_____

- -

_____

- -

_____

Materials

- ☐ seed packets of three different seed types
- ☐ adult plants (or photos) for each type of seed
- ☐ hand lens

Carry Out an Investigation

1. Look closely at seeds from three different kinds of plants.

2. Look at the adult plants or at the photos of the adult plants.

3. Match each seed with its adult plant.

Which adult plant do you think grows from each seed?

Communicate Information

1. **Record Data** Draw each seed and its matching adult plant. Use the table.

Seed	Adult Plant

2. **Draw Conclusions** From your observations, can you match the seed to the plant? Explain your answer.

_____

- -

_____

- -

_____

📓 Obtain and Communicate Information

🔤 Vocabulary

Use these words when explaining plants.

life cycle   seedling

Seeds

▶ Watch *Seeds* on how seeds become plants. Answer the questions after you have finished watching.

1. Fill in the blank.

   _____

   - - - - - - - - - - - - - -
   The _____ of a plant grow down.

2. Circle the statement that is true. Place an ✕ over the statement that is false.

   A seed might begin to grow if a bee flies by and pollinates it.

   A seed falls into the ground, and a new plant might begin to grow.

How Plants Grow

👁 Read *How Plants Grow* on the changes
a seed goes through to become a plant.

3. What does the life cycle of some
plants start with? (Circle) the answer.

   a. an adult plant

   b. a seedling

   c. a seed

4. What does a seedling grow into?

   _____

   -

   _____

Other Ways Plants Grow

👁 Read *Other Ways Plants Grow* on
plants that do not grow from seeds.

5. What is another way some plants
can make new plants?

   _____

   -

   _____

   -

   _____

✋ Inquiry Activity
Potato Plant

Materials
☐ sweet potato
☐ paper plates
☐ crayons

You will investigate how you can grow a sweet potato plant from its parts.

Make a Prediction What will happen if you leave pieces of a sweet potato on a paper plate?

_____

- -

_____

- -

_____

Carry Out an Investigation

BE CAREFUL Ask your teacher to cut the sweet potato. Follow your teacher's instructions.

1. Cut the sweet potato into pieces, with one or more "eyes" for each piece.

2. Put a sweet potato piece on a paper plate. Write your group's name on the plate.

3. Observe your sweet potato pieces every other day for two weeks.

Communicate Information

1. Record Data Draw and write your observations. Use the table.

Day	Observations

⚙ Crosscutting Concepts
Patterns

2. Communicate What happened to the sweet potato pieces? How do they help with the life cycle of the plant?

_____

- -

_____

_____

- -

_____

_____

- -

_____

⚙ Science and Engineering Practices

Complete the "I can . . ." statement.

Use examples from the lesson to explain what you can do!

_____

I can construct an

explanation

- -

_____

_____

- -

Research, Investigate, and Communicate

Inquiry Activity
Plant Life Cycle

Research You will research a plant and record the plant's life cycle.

Ask a Question What question will your research help to answer?

_____

- -

_____

Communicate Information

1. **What plant did you choose? Where does your plant live?**

_____

- -

_____

- -

2. **List the parts of your plant.**

_____

- -

_____

- -

3. Use information from your research to
draw the life cycle of your plant.

4. Write a letter or a poem to a friend.
Tell your friend about your plant.

_____

- -

_____

- -

_____

- -

_____

- -

_____

⚙ Performance Task
Life Cycle of an Apple Tree

Research You will research the life cycle of an apple tree.

Ask a Question What question will your research help to answer?

_____

_____

_____

_____

Communicate Information

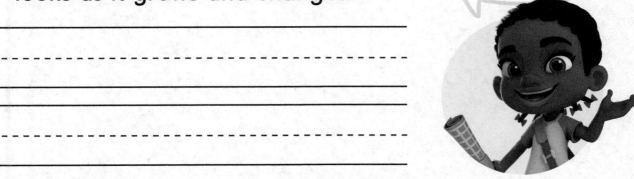

1. Describe how the apple tree looks as it grows and changes.

_____

_____

_____

_____

_____

_____

How does an apple tree grow and change?

2. Communicate Draw the life cycle of an apple tree. Show how the tree changes as it grows from a seed to an adult plant. Label the parts of the life cycle.

❓ Essential Question
How do plants grow and change?

▶ Think about the video of the plant starting to grow from a seed at the beginning of the lesson. Use what you have learned to tell how plants grow and change.

_____

- -

_____

- -

_____

- -

_____

⚙ Science and Engineering Practices

I did construct an explanation.

Rate Yourself

Color in the number of stars that tell how well you did construct an explanation.

Now that you're done with the lesson, rate how well you did.

Plants and Their Parents

PAGE KEELEY SCIENCE PROBES

Young Plants

Which friend has the best idea about young plants?

Joyce: I think young plants look exactly like their parents.

Melinda: I think young plants look like their parents but can have some differences.

Portia: I think young plants look very different from their parents.

Explain your thinking.

_____

- -

_____

- -

_____

Science in My World

Look at the photo of tulips. How are the tulips alike? How are they different? What questions do you have?

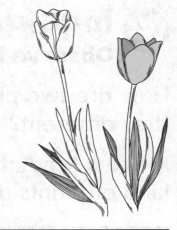

_____
- - - - - - - - - - - - - - - - - - -
_____

_____
- - - - - - - - - - - - - - - - - - -
_____

_____
- - - - - - - - - - - - - - - - - - -
_____

? **Essential Question**
How are plants like their parents?

> I love to learn about plants. I wonder how plants are like their parents.

Science and Engineering Practices

I will construct an explanation.

✋ Inquiry Activity
Observe Plants

How are two plants alike? How are they different?

Make a Prediction How are the same kind of plants different from each other?

_____

_____

_____

_____

Carry Out an Investigation

1. Take a walk outside with your class.

2. Find two plants that look the same. Observe how they are alike and different. Record your observations.

Observe the size, shape, and color of the parts of the plants!

Communicate Information

1. **Record Data** Draw the two plants you observed. Use the table.

Plant 1	Plant 2

2. **Communicate** Tell two ways the plants are alike. Tell two ways they are different.

_____

- - - - - - - - - - - - - - - - - - -

_____

_____

- - - - - - - - - - - - - - - - - - -

_____

- - - - - - - - - - - - - - - - - - -

_____

Obtain and Communicate Information

abc Vocabulary

Use these words when explaining plants and their parents.

inherit   trait

Every Plant Is Different

📖 Read *Every Plant Is Different* on how plants are similar to and different from each other. Answer the questions after you have finished reading.

1. Fill in the blank.

_____

- - - - - - - - - - - - - - - -

A young seedling can _____ some things from its parents.

2. On a single bush of roses, each flower is exactly the same. (Circle) the correct answer.

   True   False

3. Match each adult plant to its offspring.

⚙ Crosscutting Concepts
Patterns

4. How did you match the adult plant with its young plant? Explain.

_____

- -

_____

_____

- -

_____

Plant Traits

👁 Read *Plant Traits* on the way plants look and grow.

5. (Circle) the statement that is true.
   Place an ✕ on the statement that is false.

Young plants do not have the same traits as their parents.

Young plants have many of the same traits as their parents.

⚙ Science and Engineering Practices

Complete the "I can . . ." statement.

I can construct an explanation

_____

_____

_____

_____

_____

> Use examples from the lesson to explain what you can do!

🔍 Research, Investigate, and Communicate

Survival

▦ Investigate how plants grow and survive in different amounts of sunlight by conducting the simulation. Answer the questions after you have finished.

1. How did the plants grow the first year? Record what you observed. Use the table.

Sunny	Shady	Partly Sunny

2. How many of each plant did you have the second year? How many of each plant did you have after five years?

_____

- - - - - - - - - - - - - - - - - -

_____

- - - - - - - - - - - - - - - - - -

_____

- - - - - - - - - - - - - - - - - -

_____

- - - - - - - - - - - - - - - - - -

_____

3. How did the plants grow over the years? What patterns did you see? Describe what you observed.

_____

- - - - - - - - - - - - - - - - - -

_____

- - - - - - - - - - - - - - - - - -

_____

- - - - - - - - - - - - - - - - - -

_____

- - - - - - - - - - - - - - - - - -

_____

- - - - - - - - - - - - - - - - - -

_____

Performance Task
Compare Tulip Plants

You will create a model of a tulip plant and its parent.

Glue your model here.

Materials

☐ construction paper

☐ scissors

☐ glue or tape

Make a Model

BE CAREFUL Handle the scissors carefully, and follow your teacher's instructions.

1. **Research** Conduct research to find out more about tulip plants.

2. Use the materials to make a model of a young tulip plant and an adult tulip plant on a separate piece of paper.

Communicate Information

1. Make a model of your plant as a seedling. Make a model of your plant as an adult.

How are young tulip plants like their parents? How are they different?

2. Tell two ways that your young plant and adult plant are alike. Tell two ways that they are different.

_____

- -

_____

- -

_____

- -

_____

- -

_____

- -

_____

3. **Draw Conclusions** When the young plant is done growing, will it look exactly like the adult plant? Explain your answer.

- -

_____

- -

_____

- -

_____

- -

_____

? Essential Question
How are plants like their parents?

Think about the photo of tulips at the beginning of the lesson. Use what you have learned to tell how plants are like their parents.

_____

- -

_____

- -

_____

- -

_____

Science and Engineering Practices

I did construct an explanation.

Rate Yourself

Color in the number of stars that tell how well you did construct an explanation.

> Now that you're done with the lesson, rate how well you did.

Compare Animals

PAGE KEELEY SCIENCE PROBES

Comparing Animals

Which friend has the best idea about animals?

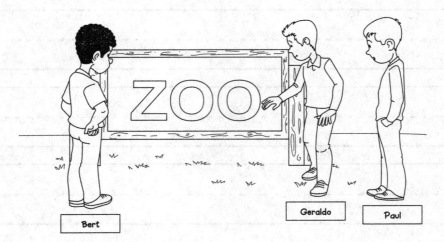

Bert: *I think animals of the same kind always look alike.*

Geraldo: *I think animals of the same kind sometimes look alike.*

Paul: *I think animals of the same kind never look alike.*

Explain your thinking.

_____

_____

_____

_____

Science in My World

Look at the photo of cows. What do you observe about the different cows? What questions do you have?

_____

- -

_____

- -

_____

- -

_____

- -

_____

? Essential Question
How are animals alike and different?

I love to learn about different animals. I wonder how the cows are alike and different.

Science and Engineering Practices

I will obtain information.
I will evaluate information.
I will communicate information.

JORDAN
Animal Trainer

✋ Inquiry Activity
Sort Animal Groups

How can you investigate ways in which animals are alike and different?

Make a Prediction What are some ways you can sort animals?

_____

- -

_____

_____

- -

_____

<table>
<tr><td colspan="2">Materials</td></tr>
<tr><td>☐</td><td>photos of many different animals</td></tr>
<tr><td>☐</td><td>scissors</td></tr>
<tr><td>☐</td><td>large paper</td></tr>
<tr><td>☐</td><td>glue</td></tr>
<tr><td>☐</td><td>markers</td></tr>
</table>

Carry Out an Investigation

BE CAREFUL Handle the scissors carefully, and follow your teacher's instructions.

1. Find photos of many different kinds of animals.

2. Cut out the photos.

3. Group the animals by ways they are similar.

4. Write labels that tell how you sorted each group.

What traits can you use to sort the animals?

Communicate Information

1. **Record Data** Use the table. Tell what traits you used to sort the animals. List the animals that were in each group.

Trait:	Trait:	Trait:

2. **Communicate** Explain why an animal was in one group instead of another.

_____

- -

_____

- -

_____

- -

_____

- -

📓 Obtain and Communicate Information

🔤 Vocabulary

Use these words when explaining animals.

mammal   bird   reptile

amphibian   fish   insect

characteristic

Types of Animals

👁 Read *Types of Animals*. Answer the question after you have finished reading.

1. Match each animal with its description.

mammal   has dry skin covered with scales

bird   lives in water and has gills

reptile   has three body parts and six legs

amphibian   gives birth to live young and takes care of them

fish   lives on land and in water

insect   has wings and feathers

Animal Characteristics

🔁 Explore the Digital Interactive *Animal Characteristics* on how the characteristics of different animals help them survive. Answer the questions after you have finished.

2. How do a butterfly's characteristics help it survive?

_____

- -

_____

_____

- -

_____

_____

- -

_____

3. Explain how cardinals are alike and different.

_____

- -

_____

_____

- -

_____

_____

- -

_____

4. Choose two animals. Think about where they live, how they move, what they look like, and what they need to survive. Fill in the Venn diagram to tell how two animals are alike and how they are different.

_____   _____
- - - - - - - - - - - - - - - -   - - - - - - - - - - - - - -
_____   _____

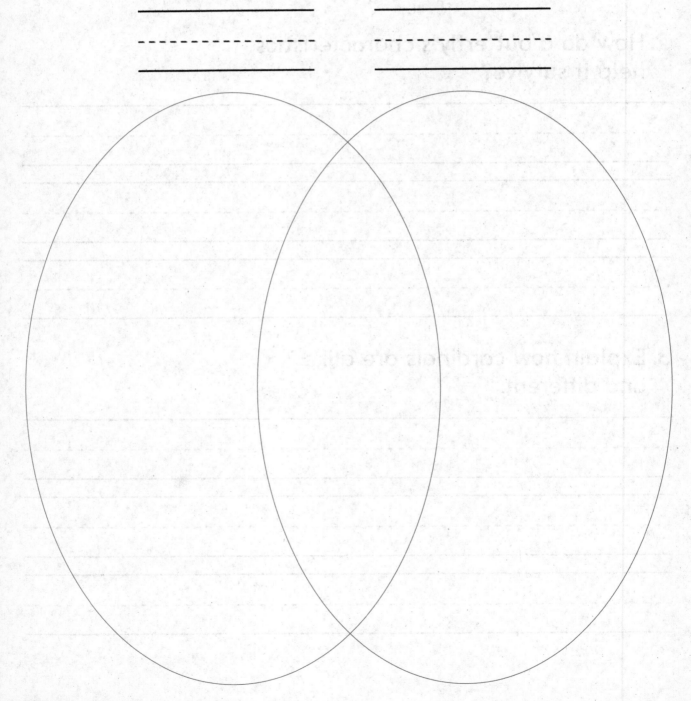

Science and Engineering Practices

Complete the "I can . . ." statements.

Use examples from the lesson to explain what you can do!

I can obtain information

I can evaluate information

I can communicate information

Research, Investigate, and Communicate

Animal Groups

▶ Watch *Animal Groups* to learn what different groups of animals are called. Answer the questions after you have finished watching.

1. What do we call a large group of fish?

   a. school

   b. flock

   c. pride

Crosscutting Concepts
Patterns

2. Does each kitten in a litter look exactly alike? Explain your answer.

_____

- -

_____

- -

_____

- -

🔩 Performance Task
Describe an Animal

Research You will research an animal and tell about it.

Ask a Question What question will your research help to answer?

_____

- -

_____

_____

- -

_____

Communicate Information

1. Choose an animal to learn more about. Draw a picture of the animal. Label what helps the animal survive.

2. Describe how one characteristic of
   your animal helps the animal survive.

_____

- -

_____

- -

_____

- -

_____

- -

_____

3. Tell how individuals of your animal are
   alike and different.

_____

- -

_____

- -

_____

- -

_____

- -

_____

? Essential Question
How are animals alike and different?

Think about the photo of cows at the beginning of the lesson. Use what you have learned to tell how the cows are alike and different.

_____

- -

_____

_____

- -

_____

⚙ Science and Engineering Practices

I did obtain information.
I did evaluate information.
I did communicate information.

Now that you're done with the lesson, rate how well you did.

Rate Yourself

Color in the number of stars that tell how well you did obtain, evaluate, and communicate information.

Animals and Their Parents

**PAGE KEELEY
SCIENCE
PROBES**

Puppies

Vinnie's dog had six puppies.
Which best describes the puppies?

☐ They look exactly like their mother.

☐ They look similar to their mother.

☐ They look very different from
their mother.

Explain your thinking.

_____

- -

_____

- -

_____

Science in My World

Look at the photo of a mother cat and her kittens. What do you observe about how the mother cat and her kittens are alike and different? What questions do you have?

_____

- -

_____

- -

_____

- -

_____

? Essential Question
How are young animals like and unlike their parents?

I love taking care of animals. I wonder how young animals are like and different from their parents!

⚙ Science and Engineering Practices

I will construct an explanation.

 Inquiry Activity
Animals

How are young animals like their parents?

Make a Prediction How are young animals and their parents alike?

_____

- -

_____

_____

- -

_____

_____

- -

_____

Make a Model

1. Think of an animal you have seen as a baby and as an adult. Draw two pictures of the animal. Show what it looks like as a baby and what it looks like as an adult.

2 Write two or three sentences to tell about your pictures.

_____

- -

_____

_____

- -

_____

_____

- -

_____

Communicate Information

1. Compare your pictures with a classmate's pictures. Use the table.

Alike	Different

📓 Obtain and Communicate Information

🔤 Vocabulary

Use these words when explaining animals and their parents.

young	carnivore	herbivore

omnivore

Families Are Similar, But Different

📖 Read pages 14–23 in *Families Are Similar, But Different.* Answer the questions.

1. Match each word with its definition.

    herbivore   eats both plants and animals

    omnivore   eats only other animals

    carnivore   eats only plants

2. How are animals of the same kind similar to each other?

    _____

    -

    _____

    -

    _____

Similarities Between Offspring and Parents

▶ Watch *Similarities Between Offspring and Parents.* Answer the questions after you have finished watching.

(Circle) True or False to describe each statement.

3. A young chicken looks very much like its parent.

   True   False

4. A grown chicken looks very much like its parent.

   True   False

Animal Life Cycle: Bird

🔊 Explore the Digital Interactive *Animal Life Cycle: Bird.* Answer the question.

5. Can you tell what the baby bird will look like as an adult? Explain.

   _____
   -
   _____
   -
   _____

Animal Life Cycle: Cow

🔄 Explore the Digital Interactive *Animal Life Cycle: Cow* on how a cow changes as it grows. Answer the questions after you have finished.

6. How is the newborn calf like its parent? How is it different?

_____
_____
_____
_____
_____
_____

7. How does the calf change during its life cycle?

_____
_____
_____
_____
_____
_____

Science and Engineering Practices

Complete the "I can . . ." statement.

I can construct an explanation

Use examples from the lesson to explain what you can do!

🔍 Research, Investigate, and Communicate

✋ Inquiry Activity
Life Cycle of a Butterfly

You will research the life cycle of a butterfly.

Make a Prediction Does a young butterfly look the same as or different from its parents?

_____

- -

_____

_____

- -

_____

Make a Model

Research Choose a type of butterfly. Research the life cycle of the butterfly.

Communicate Information

1. What type of butterfly did you research?

_____

- -

_____

2. Draw the stages in the butterfly's life cycle. Label each stage.

3. Compare your butterfly with a classmate's butterfly. How are they alike and different?

_____

- - - - - - - - - - - - - - - - - - -

_____

- - - - - - - - - - - - - - - - - - -

_____

- - - - - - - - - - - - - - - - - - -

Crosscutting Concepts
Patterns

4. Do the stages of the butterfly life cycle have the same pattern every time? Explain.

_____

- - - - - - - - - - - - - - - - - - -

_____

- - - - - - - - - - - - - - - - - - -

_____

- - - - - - - - - - - - - - - - - - -

_____

- - - - - - - - - - - - - - - - - - -

Performance Task
Compare Cat and Kittens

Materials

☐ photo of cat and kittens from Engage

You will describe ways that a parent cat and its kittens are alike and different.

Make a Prediction How are the mother cat and her kittens alike and different?

_____

- -

_____

- -

_____

- -

_____

- -

_____

Carry Out an Investigation

① Observe the photo.

② Fill in the Venn diagram. Write the word "cat" on the left side. Write the word "kittens" on the right side.

③ Write how the cat and kittens are different in the outside sections.

④ Write how the cat and kittens are alike in the middle section.

> How are the kittens like and different from their mother?

Communicate Information

1. Record your observations in your
   Venn diagram.

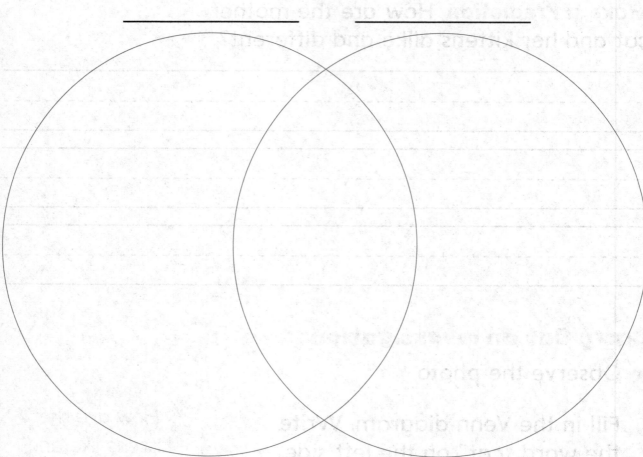

2. Draw Conclusions Are kittens exactly
   like their parents? Explain.

_____

- -

_____

_____

- -

_____

❓ Essential Question
How are young animals like and unlike their parents?

Think about the photo of a mother cat and her kittens. Use what you have learned to tell how animal young are like and unlike their parents.

_____

- -

_____

_____

- -

_____

⚙️ Science and Engineering Practices

I did construct an explanation.

Rate Yourself.

Color in the number of stars that tell how well you did construct an explanation.

Now that you're done with the lesson, rate how well you did.

Offspring and Survival

PAGE KEELEY SCIENCE PROBES

Baby Animals
Who has the best idea about how baby animals survive?

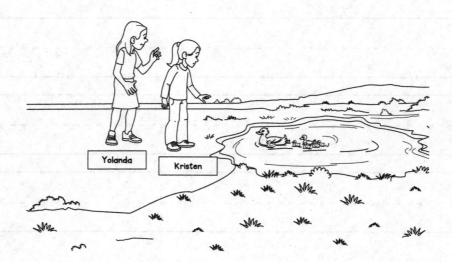

Yolanda: I think the ducklings learn how to survive from their parents and each other.

Kristen: I think the ducklings are born knowing how to survive.

Explain your thinking.

_____

- -

_____

- -

_____

Science in My World

Look at the photo of birds. Listen to the sound of baby birds chirping. What questions do you have?

_____

- -

_____

_____

- -

_____

_____

- -

_____

? Essential Question

How do animal offspring survive?

> I want to learn more about animals. I wonder how animals care for their young.

Science and Engineering Practices

I will obtain information.
I will evaluate information.
I will communicate information.

✋ Inquiry Activity
Animal Young

Materials

☐ magazines

☐ scissors

☑ paper

☐ glue

How do parents protect their young?

Make a Prediction What are some ways animals protect their young?

_____

- -

- -

- -

- -

- -

Carry Out an Investigation

BE CAREFUL Handle the scissors carefully, and follow your teacher's instructions.

1. Look through magazines to find photos of animals with their offspring.

2. Cut out the photos. Glue the photos on a separate piece of paper.

How do animals protect their young and teach them how to survive?

Communicate Information

Look at one of your photos.
Complete the sentences.

1. My photo is of a/an

_____

- - - - - - - - - - - - - - - - - - -

_____

- - - - - - - - - - - - - - - - - - -

_____

2. The adult is

_____

- - - - - - - - - - - - - - - - - - -

_____

- - - - - - - - - - - - - - - - - - -

_____

3. The adult helps the young

_____

- - - - - - - - - - - - - - - - - - -

_____

- - - - - - - - - - - - - - - - - - -

_____

- - - - - - - - - - - - - - - - - - -

_____

Glue your paper with photos here.

 Obtain and Communicate Information

🔤 Vocabulary

> Use these words when explaining how animal offspring survive.
>
> behavior   learn   signal
>
> adaptation

Animal Communication

▶ Watch *Animal Communication* on ways that baby animals tell their parents what they need.

1. Tell one way that baby animals let their parents know they are ready to eat.

_____

_____

_____

_____

Animal Messages

📖 Read *Animal Messages* on how adult animals help their offspring.

2. Some animals are born with behaviors they need to survive and do not need to learn from their parents. (Circle) the correct answer.

   True   False

Fill in the blanks to complete the sentences.

3. A chick _____
   when it is hungry.

4. A baby bat gives off a _____
   when it is ready for milk.

5. A dolphin calf _____
   to communicate with its mother.

Animals Meet Their Needs

Explore the Digital Interactive *Animals Meet Their Needs*. Answer the questions after you have finished.

⚙ Crosscutting Concepts
Patterns

6. How do baby animals get what they need to survive?

_____

- -

_____

- -

_____

- -

_____

- -

_____

- -

_____

- -

_____

- -

Science and Engineering Practices

Complete the "I can . . ." statements.

Use examples from the lesson to explain what you can do!

I can obtain information

I can evaluate information

I can communicate information

Research, Investigate, and Communicate

Needs of Living Things

▶ Watch *Needs of Living Things*.
Answer the questions after you have finished watching.

1. What is the difference between a carnivore and an herbivore?

_____

- -

_____

_____

- -

_____

2. Draw an example of a carnivore.
Draw an example of an herbivore.

Carnivore	Herbivore

✋ Inquiry Activity
Animal Teeth

You will make models of carnivore and herbivore teeth.

Materials
☐ modeling clay

Make a Prediction How are carnivore teeth different from herbivore teeth?

_____

- -

_____

- -

_____

How do different types of teeth help animals get what they need to survive?

Make a Model

Research animal teeth to find out what carnivore teeth and herbivore teeth look like. Use the clay to make a model of each. Draw a picture of your models.

Carnivore Teeth	Herbivore Teeth

Communicate Information

1. Describe the two types of teeth.
   How are they alike or different?

----

2. **Draw Conclusions** How do carnivores'
   and herbivores' teeth help them get
   what they need to survive?

Performance Task
Young Animal Book

Research You will research an animal and explain how the animal helps its offspring survive.

Ask a Question What question will your research help to answer?

_____

- -

_____

- -

_____

Communicate Information

1. Draw a picture of your animal helping its young.

2. Make a short book that tells how your animal helps its offspring survive. You may add pictures to your book.

3. Tell the main idea and key details you included in your book.

_____

_____

_____

_____

_____

_____

4. How did you research information about your animal? Did your sources help you to write your book? Why or why not?

_____

_____

_____

_____

_____

_____

❓ Essential Question
How do animal offspring survive?

Think about the photo of birds at the beginning of the lesson. Use what you have learned to explain how animal offspring survive.

_____

- -

_____

_____

_____

⚙️ Science and Engineering Practices

I did obtain information.
I did evaluate information.
I did communicate information.

Now that you're done with the lesson, rate how well you did.

Rate Yourself

Color in the number of stars that tell how well you did obtain, evaluate, and communicate information.

Offspring and Their Parents

⚙ Performance Project
Bean Plant Life Cycle

You will investigate the life cycle of bean plants.

Make a Prediction How does a bean plant grow and change during its life cycle?

_____

- -

_____

_____

- -

_____

Carry Out an Investigation

List the materials you will need and the steps you will follow.

_____

- -

_____

_____

- -

_____

All living things grow and change during their life cycle!

Communicate Information

Write or draw your observations.

🌐 Explore More in My World

Did you learn the answers to all of your questions from the beginning of the module? If not, how could you design an experiment or conduct research to help answer them?

Earth and Space

Science in My World

Have you ever looked at the night sky?
What did you see? Did you see the moon
and stars? What questions do you have
about objects in the night sky?

_____

- -

_____

_____

- -

_____

🔤 Key Vocabulary

Look and listen for these words as you
learn about Earth and space.

daytime	Moon	nighttime
phases	planet	position
rotate	season	star
Sun		

How can you predict patterns in the sky?

EMILY
Aerospace Engineer

Emily wants to be an aerospace engineer. An aerospace engineer designs spaceships and other vehicles that help people explore our solar system. Emily is curious about the night sky. What objects do you see in the night sky where you live? Draw what you see in the sky at night.

Science and Engineering Practices

I will analyze data.
I will interpret data.

Day and Night

 PAGE KEELEY SCIENCE PROBES

Day and Night

Which friend has the best idea about why we have day and night?

| Maribel | Benito | Hoshi | Dave |

Dave: *I think it is because Earth spins all the way around each day.*

Benito: *I think it is because Earth goes around the Sun once each day.*

Hoshi: *I think it is because the Sun goes around Earth once each day.*

Maribel: *I think it is because the Sun goes up and down once each day.*

Explain your thinking.

_____

- -

_____

- -

_____

Science in My World

Look at the photo of the Sun. What do you observe about the Sun and the sky? What questions do you have?

? Essential Question
What causes the pattern of day and night?

> I love to look at the sky. I wonder why the Sun seems to be in different places during the day!

HUGO
Meteorologist

🪃 Science and Engineering Practices

I will analyze data.
I will interpret data.

Inquiry Activity
Shadows

How does light affect the length and direction of shadows?

Make a Prediction What causes shadows to change?

_____

- -

_____

_____

- -

_____

Materials

- [ ] craft sticks
- [ ] modeling clay
- [ ] paper
- [ ] flashlight
- [ ] pencil

Carry Out an Investigation

BE CAREFUL Wash your hands before and after using modeling clay.

1. Put a craft stick in clay on a piece of paper.

2. Shine a light on the craft stick.

3. Trace the shadow and label it.

4. Move the flashlight to another spot.

5. Trace and label the new shadow.

Observe what happens to the shadow when you move the flashlight.

Communicate Information

1. Record Data Draw the light, stick, and shadows. Use the table.

	Shadows
First Shadow	
Second Shadow	

2. Communicate How did you make a long shadow?

_____

- - - - - - - - - - - - - - - - - - - -

_____

_____

- - - - - - - - - - - - - - - - - - - -

_____

3. How did you make a short shadow?

_____

- - - - - - - - - - - - - - - - - - - -

_____

_____

- - - - - - - - - - - - - - - - - - - -

_____

4. Draw Conclusions What caused the shadows to change?

_____

- - - - - - - - - - - - - - - - - - - -

_____

_____

- - - - - - - - - - - - - - - - - - - -

_____

_____

- - - - - - - - - - - - - - - - - - - -

_____

💬 Obtain and Communicate Information

🔤 Vocabulary

Use these words when explaining day and night.

Sun	daytime	nighttime
rotate	planet	

Day and Night

▶ Watch *Day and Night* on the rotation of Earth around the Sun.

1. When Earth spins around, we say it

   _____

   -

   _____ .

2. (Circle) the statement that is true.
   Place an ✕ over the statement that is false.

   It is daytime when your home faces the Sun.

   It is nighttime when your home faces the Sun.

✋ Inquiry Activity
The Sun and Earth

Materials

☐ stickers or sticky paper

☐ globe

☐ flashlight

You will investigate how Earth's movement causes day and night.

Make a Prediction How does Earth's movement cause day and night?

_____

- -

_____

- -

_____

- -

_____

Carry Out an Investigation

BE CAREFUL Follow your teacher's instructions.

1. Put a sticker or sticky paper where you live on the globe.

2. Shine a light on the globe.

3. Spin the globe around. Let it stop by itself.

4. Observe where the light is shining on the globe.

The light from the flashlight shining on the globe is a model of the light from the Sun shining on Earth.

Communicate Information

3. Communicate Is it daytime or nighttime where you live? How do you know?

_____

- -

_____

_____

- -

_____

4. Is it daytime or nighttime on the other side of Earth? How do you know?

_____

- -

_____

_____

- -

_____

Earth's Sky Changes

📖 Read pages 14–23 in *Earth's Sky Changes.*

5. Fill in the blank. Use a vocabulary word.

_____

- - - - - - - - - - - - - - - - -

Earth is a _____ that revolves or moves around the Sun.

6. Why do we see different objects in
the sky at different times of day?

_____
- -
_____
_____
- -
_____

7. The Sun does not move across the sky.
It looks like it does because Earth
is moving. (Circle) the correct answer.

   True   False

⚙ Crosscutting Concepts
Patterns

8. What causes the pattern of day
and night?

_____
- -
_____
_____
- -
_____
_____
- -
_____

Science and Engineering Practices

Complete the "I can . . ." statements.

I can analyze data

I can interpret data

Use examples from the lesson to explain what you can do!

Research, Investigate, and Communicate

✋ Inquiry Activity
Measuring Your Shadow

Materials
☐ meterstick or tape measure
☐ chalk

You will investigate how the length and direction of your shadow changes during the day.

Make a Prediction How will your shadow change during the day?

_____

_____

_____

_____

Carry Out an Investigation

BE CAREFUL Do not look directly at the Sun. Follow your teacher's instructions.

1. Measure your shadow in the morning, at noon, and in the afternoon.

2. Record your measurements.

3. Record where the Sun is in the sky.

Communicate Information

1. **Record Data** Record the length of your shadow. Show where the Sun is in the sky. Use the table.

Your Shadow During the Day		
Morning	**Noon**	**Afternoon**

2. **Communicate** How did your shadow change?

_____

_____

_____

3. What pattern do you observe about your shadow and where the Sun is in the sky?

_____

- -

_____

_____

- -

_____

_____

- -

_____

4. Make a list of your three shadows by length. Start with the shortest.

_____

- -

_____

_____

- -

_____

_____

- -

_____

⚙ Performance Task
The Sun During the Day

You will record where the Sun is at different times of the day.

Materials
- [ ] drawing paper
- [ ] crayons
- [ ] pencil

Make a Prediction Where is the Sun during different times of the day?

_____

- -

_____

_____

- -

_____

Make a Model

BE CAREFUL Do not look directly at the Sun. Follow your teacher's instructions.

1. Draw lines to divide your paper into three sections. Write "Morning" at the top of one section. Write "Noon" at the top of the next section, and write "Afternoon" at the top of the last section.

2. Go outside in the morning. Draw a picture of the Sun in the sky. Be sure to show land in your picture.

3. Repeat step 2 at noon and in the afternoon.

Communicate Information

1. **Communicate** Compare your pictures. What does it look like the Sun is doing?

_____

- -

_____

- -

_____

2. Think back to what you learned about how Earth moves. Why does it look like the Sun moves across the sky during the day?

_____

- -

_____

- -

_____

3. **Draw Conclusions** Where will the Sun be in the morning, at noon, and in the afternoon tomorrow?

_____

- -

_____

- -

_____

❓ Essential Question
What causes the pattern of day and night?

Think about the photo of the Sun at the beginning of the lesson. Use what you have learned to tell what causes day and night.

_____

- -

_____

_____

- -

_____

⚙️ Science and Engineering Practices

I did analyze data.
I did interpret data.

Now that you're done with the lesson, rate how well you did.

Rate Yourself

Color in the number of stars that tell how well you did analyze and interpret data.

Seasonal Patterns

PAGE KEELEY
SCIENCE
PROBES

Daylight Hours

Which friend has the best idea about hours of daylight?

Carmen   Matt   Vickie   Jason

Jason: I think daylight is longest in the spring.

Vickie: I think daylight is longest in the summer.

Matt: I think daylight is longest in the fall.

Carmen: I think daylight is longest in the winter.

Explain your thinking.

_____

- -

_____

- -

_____

Science in My World

▶ Look at the video of the trees. What happens to the leaves on the trees? What questions do you have?

_____

- -

_____

- -

_____

- -

_____

- -

_____

? Essential Question
What causes the seasons?

I love learning about weather. I wonder how the weather changes throughout the year!

⚙ Science and Engineering Practices

I will plan an investigation.
I will carry out an investigation.

Inquiry Activity
Sunlight

▦ Conduct the simulation to investigate seasonal changes to sunrise, sunset, and weather in different places around the world. Fill in the table and answer the questions after you have finished.

1. How is weather different around the world? Fill in the table.

	February Temperature	August Temperature
Fairbanks, Alaska		
Lebanon, Kansas		
Macapa, Brazil		
Buenos Aires, Argentina		

2. Do all places around the world have the same weather at the same time of year?

3. Which two cities are coldest in August?

4. Which city has the warmest February?

5. Does Lebanon or Macapa have fewer hours of daylight in November?

_____

- -

_____

- -

_____

6. Watch the sunrises and sunsets in Buenos Aires during the year. How does the height of the Sun in the sky change from January to December?

_____

- -

_____

- -

_____

- -

_____

- -

_____

- -

_____

- -

_____

💬 Obtain and Communicate Information

🔤 Vocabulary

Use these words when explaining seasonal patterns.

season   spring   summer

fall   winter

Seasons Change

▶ Watch *Seasons Change* on the different seasons. Answer the question after you have finished watching.

1. What is one thing you might do in spring? (Circle) the answer.

   a. You might build a snowman.

   b. You might fly kites.

   c. You might rake colored leaves.

The Four Seasons

📖 Read pages 14–23 in *The Four Seasons.*
Answer the questions after you have
finished reading.

2. The part of Earth that tilts toward the
Sun has the season of

_____

- - - - - - - - - - - - - - - - - - - -

_____.

3. Spring days have more hours of
daylight than winter days. (Circle)
the correct answer.

   True   False

⚙ Crosscutting Concepts
Patterns

4. What is the pattern of the number
of daylight hours during the different
seasons?

_____

- -

_____

- -

_____

Science and Engineering Practices

Complete the "I can . . ." statements.

I can plan an investigation

I can carry out an investigation

Use examples from the lesson to explain what you can do!

Research, Investigate, and Communicate

How Earth Moves

Explore the Digital Interactive *How Earth Moves* on Earth's tilt, its movement around the Sun, and seasons. Answer the question after you have finished.

1. Match each season with the sentence that best describes it. Draw a line.

spring   North America has the least amount of daylight.

summer   North America has more daylight than in winter.

fall   North America has the most daylight.

winter   North America has less daylight than in summer.

Look at the Facts

Use the tables to answer questions about the seasons.

Table A

Sunrise	Sunset
7:02 A.M.	5:15 P.M.
7:03 A.M.	5:14 P.M.
7:04 A.M.	5:13 P.M.

Table B

Sunrise	Sunset
6:07 A.M.	9:05 P.M.
6:08 A.M.	9:04 P.M.
6:09 A.M.	9:03 P.M.

2. Which table shows sunrise and sunset times for summer? How do you know?

_____

- -

_____

_____

- -

_____

_____

- -

_____

3. Which table shows sunrise and sunset times for winter? How do you know?

_____
- -
_____
- -
_____
- -
_____
- -

4. Predict what season comes after Table B. Explain.

_____
- -
_____
- -
_____
- -
_____
- -

⚙ Performance Task
How Some Trees Change

You will use what you have learned to show how some trees change at different times of the year.

Materials
☐ pencil
☐ crayons

1. Draw the same tree for each season. Show how the tree changes.

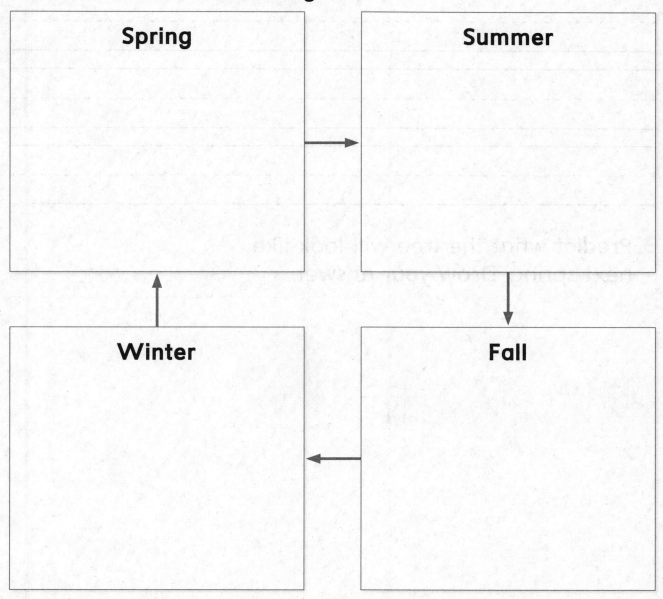

Spring → Summer

Winter ↑   Fall ←

2. How much daylight does the tree get in each season?

_____

- -

_____

- -

_____

- -

_____

- -

_____

- -

_____

3. Predict what the tree will look like next spring. Draw your answer.

❓ Essential Question
What causes the seasons?

▶ Think about the video of the trees from the beginning of the lesson. Use what you have learned to explain what season it is.

_____

- -

_____

_____

- -

_____

- -

_____

⚙️ Science and Engineering Practices

I did plan an investigation.
I did carry out an investigation.

Rate Yourself

Color in the number of stars that tell how well you did plan and carry out an investigation.

Now that you're done with the lesson, rate how well you did.

The Moon

PAGE KEELEY
SCIENCE
PROBES

Moon Patterns

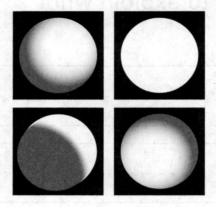

The part of the Moon we see changes.
These changes are called moon phases.
How long does the pattern of the Moon
phases take? Circle your prediction.

A. about 1 night

B. about 8 days

C. about 4 weeks

D. about 1 year

Explain your thinking.

_____

- -

_____

- -

_____

Science in My World

Look at the photo of the moon. What do you observe about the moon? What questions do you have?

_____

- - - - - - - - - - - - - - - - - - -

_____

- - - - - - - - - - - - - - - - - - -

_____

- - - - - - - - - - - - - - - - - - -

_____

- - - - - - - - - - - - - - - - - - -

? Essential Question
Why does the Moon seem to change?

⚙ Science and Engineering Practices

I will analyze data.
I will interpret data.

I love to look at objects in the night sky. I wonder why the Moon seems to change!

HALEY
Astronomer

🖐 Inquiry Activity
How the Moon Looks

What does the Moon look like in the night sky?

Make a Prediction What shape will the Moon be in the night sky?

_____

- - - - - - - - - - - - - - - - - - - -

_____

_____

- - - - - - - - - - - - - - - - - - - -

_____

- - - - - - - - - - - - - - - - - - - -

_____

Carry Out an Investigation

BE CAREFUL Observe the Moon outside at night with an adult.

① Look at the Moon outside on a clear night.

② Draw a picture of what you observe.

③ Take your picture to school. Compare it with the pictures of your classmates.

What is the shape of the Moon?

Communicate Information

1. **Record Data** Draw what the Moon looked like.

2. **Communicate** How did your drawing compare to your classmates' drawings?

_____

- -

_____

- -

_____

- -

_____

 Obtain and Communicate Information

🔤 Vocabulary

Use these words when explaining the moon.

Moon   phases   position

The Moon

▶ Watch *The Moon* on how the Moon moves around Earth. (Circle) the correct answers

1. We can see only the part of the Moon that is lit up by the Sun.

   True   False

2. When the Moon is between Earth and the Sun, we can see the whole moon.

   True   False

3. Earth moves around the Sun, but the Moon stays in one place.

   True   False

✋ Inquiry Activity
Make a Model

You will make a model to show the pattern of how the Moon changes.

Make a Prediction How does the shape of the Moon change?

Materials

☐ photos of the Moon in different phases

☐ pencil

☐ paper

_____

- -

_____

- -

_____

- -

Glue your drawing here.

Make a Model

1. Observe the phase of the Moon in the photo.

2. Draw the next three phases of the Moon on a separate piece of paper.

3. Label the phases. Use arrows to show the order of the phases.

Communicate Information

4. **Draw Conclusions** Describe the position of Earth and the Moon during the first phase in your drawing.

⚙ Crosscutting Concepts
Patterns

5. How did you know what the Moon would look like in the next three phases?

_____
- -
_____
_____
- -
_____
_____
- -
_____

Science and Engineering Practices

Complete the "I can . . ." statements.

I can analyze data

I can interpret data

Use examples from the lesson to explain what you can do!

Research, Investigate, and Communicate

Inquiry Activity
Moon Observations

You will investigate how the Moon changes during the month.

Make a Prediction How will the Moon change during the month?

_____

- -

_____

- -

_____

Carry Out an Investigation

BE CAREFUL Follow your teacher's instructions. Observe the Moon outside at night with an adult.

1. Look at the Moon each night for one month.

2. When you have completed your moon observations, use your data to make a book to show the phases of the Moon.

Materials
- [ ] paper
- [ ] pencil

Communicate Information

1. Record Data Draw a picture of your observations each night. Label each drawing with the date.

2. Draw Conclusions Why do the phases of the Moon occur in the same order every month?

_____

_____

_____

_____

3. How is the position of the Moon related to the phases of the Moon?

_____

_____

_____

_____

Performance Task
Phases of the Moon

You will make a model to show the phases of the Moon.

Make a Model

1 Draw the shape of the Moon in each phase on white paper.

2 Cut out the shapes.

3 Put the shapes in the correct order to show the phases of the Moon. Glue the shapes onto black construction paper.

4 Use the white crayon to add labels and arrows. Identify the phases of the Moon and their correct order.

Materials

☐ pencil

☐ white paper or construction paper

☐ scissors

☐ glue

☐ black construction paper

☐ white crayon

How can you model the phases of the Moon?

Communicate Information

1. List the order of the phases of
the Moon on your model.

_____

_____

_____

_____

_____

_____

_____

_____

2. How did you decide the correct order
for your pictures?

_____

_____

_____

_____

_____

_____

_____

_____

? Essential Question

Why does the Moon seem to change?

Think about the photo of the Moon at the beginning of the lesson. Use what you have learned to explain why the Moon seems to change.

_____

- -

_____

- -

_____

- -

_____

Science and Engineering Practices

I did analyze data.
I did interpret data.

Now that you're done with the lesson, rate how well you did.

Rate Yourself

Color in the number of stars that tell how well you did analyze and interpret data.

The Sun and Stars

PAGE KEELEY SCIENCE PROBES

Seeing Stars

Which friend has the best idea about when we see stars?

Lucy   Elmer   Akira

Elmer: *I think we see stars only in the daytime.*

Lucy: *I think we see stars only in the nighttime.*

Akira: *I think we see stars in the daytime and in the nighttime.*

_____

- -

_____

- -

_____

Science in My World

Look at the photo of the night sky. What do you wonder about the objects that you see in the night sky? What questions do you have?

_____

- -

_____

- -

_____

- -

_____

- -

_____

? Essential Question

How can you describe the Sun and stars?

> I love to look at the sky during the day and at night. I wonder what the Sun and other stars are like!

⚙ Science and Engineering Practices

I will analyze data.
I will interpret data.

✋ Inquiry Activity
Record Data

Materials

☐ pencil

☐ thermometer

☐ red crayon

☐ blue crayon

How do the position of the Sun and the temperature change during the day?

Make a Prediction What time of day will be the warmest?

_____

- -

_____

_____

- -

_____

- -

_____

Carry Out an Investigation

BE CAREFUL Follow your teacher's directions. Do not look directly at the Sun.

1 Observe the Sun and the temperature throughout the day.

2 Record the temperature at different times of the day.

3 Repeat your observations at the same times for two more days.

Communicate Information

1. Record Data How does the temperature change throughout the day? Record the temperature. Use the table.

Sun Location and Time of Day	Day 1 Temperature	Day 2 Temperature	Day 3 Temperature
Sun Rising Before School			
Sun High in Sky at Lunchtime			
Sun Setting After Dinner			

2. Find the highest temperature for each day. (Circle) it in red. Find the lowest temperature for each day. (Circle) it in blue.

3. **Communicate** When was the temperature warmest each day?

_____

- -

_____

- -

_____

4. Describe how the position of the Sun and the temperature changed each day. Was it the same all three days you observed?

_____

- -

_____

- -

_____

- -

_____

- -

_____

- -

_____

- -

_____

💬 Obtain and Communicate Information

🔤 Vocabulary

> Use this word when explaining the Sun and stars.
>
> star

The Sun and Stars

▶ Watch *The Sun and Stars* on objects found in the sky. Answer the question.

1. (Circle) the statement that is true. Place an ✕ over the statement that is false.

   The Sun is a planet.

   Stars make their own light.

Lights in the Sky

📖 Read pages 14–23 in *Lights in the Sky*.

2. The Sun gives off energy in the form of

   _____   _____

   -

   _____ and _____.

3. Why is the Sun important to Earth?

_____

- -

_____

_____

- -

_____

4. When do stars shine?
(Circle) the correct answer.

a. During the day

b. During the night

c. All the time

The Sun in the Sky

📲 Explore the Digital Interactive *The Sun in the Sky* on how the Sun moves across the sky during the day.

5. Place a 1, 2, or 3 in each box to show the correct order of how the Sun moves across the sky during the day.

☐ The Sun is high in the sky.

☐ The Sun goes below the trees.

☐ The Sun begins to rise above the buildings.

Science and Engineering Practices

Complete the "I can . . ." statements.

I can analyze data

I can interpret data

Use examples from the lesson to explain what you can do!

Research, Investigate, and Communicate

Materials
- [ ] paper
- [ ] pencil

Inquiry Activity
Near and Far

You will investigate how objects look different when they are near or far.

Make a Prediction How does an object's distance change how big or small it looks?

_____

- -

_____

- -

_____

Carry Out an Investigation

1. Work with a partner. Draw a picture on the next page of your partner near an object.

2. Draw another picture after your partner moves far away from the object.

How does your partner's size seem to change?

Communicate Information

1. **Record Data** Draw your pictures in the space below. Include your partner and the object you observed.

Classmate Near the Object

Classmate Far from the Object

2. **Communicate** Does your classmate appear to be the same size in both pictures? Tell how the pictures are alike and different.

⚙ Crosscutting Concepts
Patterns

3. Relate what you just learned about being near and far from an object. The Sun is a star. Why do other stars look very small?

⚙ Performance Task
Observe the Night Sky

Materials
☐ paper
☐ pencil

You will observe objects in the night sky.

Make a Prediction What objects can you see in the night sky?

_____

- -

_____

_____

- -

_____

Carry Out an Investigation

BE CAREFUL Follow your teacher's instructions. Observe the night sky outside with an adult.

1. Observe the sky outside on a clear night.

2. Draw a picture of what you see to record your observations.

3. Compare your observations with a classmate's observations.

Be like an astronomer! What do you see in the night sky?

Communicate Information

1. Record Data Draw what you observed.
   Label the objects in your drawing.

2. Compare your drawing with
   a classmate's drawing. Did you see
   the same things? Tell how your
   drawings are alike and different.

_____

- -

_____

- -

_____

- -

_____

- -

_____

? Essential Question
How can you describe the Sun and stars?

Think about the photo of the night sky at the beginning of the lesson. Use what you have learned to describe the Sun and stars.

_____

- -

_____

- -

_____

Science and Engineering Practices

I did analyze data.
I did interpret data.

Now that you're done with the lesson, rate how well you did.

Rate Yourself

Color in the number of stars that tell how well you did analyze and interpret data.

Earth and Space

⚙ Performance Project
Observing the Moon

You will observe the position of the Moon during the night.

Make a Prediction How does the Moon change throughout the night?

_____

- -

_____

Carry Out an Investigation

BE CAREFUL Follow your teacher's instructions. Observe the night sky outside with an adult.

1. Observe the Moon on a clear night.

2. Make observations every 30 minutes for two hours.

3. **Record Data** Make a table to organize and draw your observations. Label each drawing with the time you made the observation.

Patterns of the Sun, Moon, and stars can be observed, described, and predicted.

Communicate Information

Draw Conclusions What will happen to the Moon one hour after your last observation? What will happen in two hours? Describe how the Moon will change.

_____
- -
_____
_____
- -
_____
_____
- -
_____
_____
- -
_____

 Explore More in My World

Did you learn the answers to all of your questions from the beginning of the module? If not, how could you design an experiment or conduct research to help answer them?

Dinah Zike Explaining
Visual Kinesthetic Vocabulary®, or VKVs®

What are VKVs and who needs them?

> VKVs are flashcards that animate words by kinesthetically focusing on their structure, use, and meaning. VKVs are beneficial not only to students learning the specialized vocabulary of a content area, but also to students learning the vocabulary of a second language.

Dinah Zike | Educational Consultant
Dinah-Might Activities, Inc. — San Antonio, Texas

Why did you invent VKVs?

> Twenty years ago, I began designing flashcards that would accomplish the same thing with academic vocabulary and cognates that Foldables® do with general information, concepts, and ideas—make them a visual, kinesthetic, and memorable experience.

I had three goals in mind:

- **Making two-dimensional flashcards three-dimensional**

- **Designing flashcards that allow one or more parts of a word or phrase to be manipulated and changed to form numerous terms based upon a commonality**

- **Using one sheet or strip of paper to make purposefully shaped flashcards that were neither glued nor stapled, but could be folded to the same height, making them easy to stack and store**

Why are VKVs important in today's classroom?

> At the beginning of this century, research and reports indicated the importance of vocabulary to overall academic achievement. This research resulted in a more comprehensive teaching of academic vocabulary and a focus on the use of cognates to help students learn a second language. Teachers know the importance of using a variety of strategies to teach vocabulary to a diverse population of students. VKVs function as one of those strategies.

VKV286

An Interview with
Dinah Zike Explaining
Visual Kinesthetic Vocabulary®, or VKVs®

Dinah Zike's
Visual
Kinesthetic
Vocabulary®

How are VKVs used to teach content vocabulary?

As an example, let's look at content terms based upon the combining form *–vore*. Within a unit of study, students might use a VKV to kinesthetically and visually interact with the terms *herbivore*, *carnivore*, and *omnivore*. Students note that *–vore* is common to all three words and it means "one that eats" meat, plants, or both depending on the root word that precedes it on the VKV. When the term *insectivore* is introduced in a classroom discussion, students have a foundation for understanding the term based upon their VKV experiences. And hopefully, if students encounter the term *frugivore* at some point in their future, they will still relate the *–vore* to diet, and possibly use the context of the word's use to determine it relates to a diet of fruit.

What organization and usage hints would you give teachers using VKVs?

Cut off the flap of a 6" x 9" envelope and slightly widen the envelope's opening by cutting away a shallow V or half circle on one side only. Glue the non-cut side of the envelope into the front or back of student workbooks or journals. VKVs can be stored in the pocket.

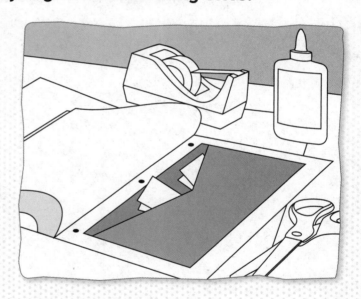

Encourage students to individualize their flashcards by writing notes, sketching diagrams, recording examples, forming plurals (radius: radii or radiuses), and noting when the math terms presented are homophones (sine/sign) or contain root words or combining forms (kilo-, milli-, tri-).

As students make and use the flashcards included in this text, they will learn how to design their own VKVs. Provide time for students to design, create, and share their flashcards with classmates.

Dinah Zike's book Foldables, Notebook Foldables, & VKVs for Spelling and Vocabulary 4th-12th won a Teachers' Choice Award in 2011 for "instructional value, ease of use, quality, and innovation"; it has become a popular methods resource for teaching and learning vocabulary.

Copyright © McGraw-Hill Education.

states of matter

States of matter include:

A _____ is the quick back and forth movement of an object.

_____ means to move back and forth quickly.

vibrate

Dinah Zike's
Visual
Kinesthetic
Vocabulary ®

Sound Energy

✂ cut on all dashed lines

fold on all solid lines

ion

made of.

_____

is what all things are

Memory Maker: Draw something that vibrates.

Memory Maker: Draw water in different states of matter.

Copyright © McGraw-Hill Education.

Copyright © McGraw-Hill Education.

VKV

Dinah Zike's
Visual
Kinesthetic
Vocabulary®

✂ cut on all dashed lines

▱ fold on all solid lines

transparent

Something is _____ when light cannot pass through.

shades do not let sunlight into a room.

Something is _____ when some light can pass through.

Sunglasses are _____

Something is _____ when all light can pass through.

Something is _____ when all light can pass through.

What's another word for *transparent*? (Circle one.)
A. *clear*
B. *cloudy*
C. *muddy*

Memory Maker: List or draw examples of the words on this VKV.

opaque

translucent

transparent

What's another word for *transparent*? (Circle one.)

A. *clear*

B. *cloudy*

C. *muddy*

lucent

Copyright © McGraw-Hill Education.

Copyright © McGraw-Hill Education.

Dinah Zike's
Visual
Kinesthetic
Vocabulary®

cut on all dashed lines

fold on all solid lines

A _____ is a dark shape that is created when a source of light is blocked.

Which would make a shadow, a clear glass of water or a mug of cocoa? Circle one.

_____ is a form of energy that lets you see.

Draw examples of sources of light.

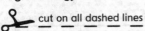
shadow

light

Memory Maker: Draw how a shadow is made on a sunny day.

Copyright © McGraw-Hill Education.

VKV

Dinah Zike's
Visual
Kinesthetic
Vocabulary®

Use Energy to Communicate

✂ cut on all dashed lines   ⬜ fold on all solid lines

electric

Something is electric if it is powered by _____ .

_____ is a form of energy that gives some things the power to work.

To _____ means to give back an image.

A _____ is a smooth piece of glass that reflect what is in front of it.

Mirrors _____ images of what is in front of them.

A reflection of yourself is what you see when you look in a _____ .

Copyright © McGraw-Hill Education.

REFLECT

Memory Maker: Write what you see when you hold the word *reflect* up to a mirror.

reflect

mirror

ity

Memory Maker: Draw examples of the words on this VKV.

Copyright © McGraw-Hill Education.

Copyright © McGraw-Hill Education.

VKV

Dinah Zike's
Visual
Kinesthetic
Vocabulary®

Plant and Animal Parts

✂ cut on all dashed lines

▭ fold on all solid lines

A _____ is an animal that lives in water and has gills and fins.

An _____ is an animal that lives part of its life in water and part on land.

A _____ is an animal that has dry skin covered with scales.

fish

amphibian

reptile

Dinah Zike's
VKV Visual
Kinesthetic
Vocabulary®

Plant and Animal Parts

✂ cut on all dashed lines   🗋 fold on all solid lines

S

S

Memory Maker: Draw examples of reptiles.

Memory Maker: Draw amphibians. Be sure to include details about where they live.

Memory Maker: Draw a detailed fish. Include words from this VKV.

Copyright © McGraw-Hill Education.

Copyright © McGraw-Hill Education.

A _____ is an animal that has two legs, two wings, and feathers.

A _____ is the plant part that holds the seeds.

An _____ is an animal with three body sections and six legs.

bird

fruit

insect

Dinah Zike's
VKV Visual
Kinesthetic
Vocabulary®

S

S

S

Memory Maker: Draw a bird that you have seen near your home.

Memory Maker: Make a word web of examples of the words on this VKV. You may draw if needed.

Memory Maker: Draw an example of an insect.

Copyright © McGraw-Hill Education.

Copyright © McGraw-Hill Education.

VKV Dinah Zike's
Visual
Kinesthetic
Vocabulary®

Plant and Animal Parts

✂ cut on all dashed lines

▭ fold on all solid lines

mammal

non-living

A _____
is an animal with
hair or fur.

_____ thing is a
A _____
thing that does not grow and change,
or need food, air, or water to survive.

Dinah Zike's
VKV Visual
Kinesthetic
Vocabulary®

Plant and Animal Parts

✂ cut on all dashed lines   📄▸ fold on all solid lines

Memory Maker: Look around you. Find and draw a nonliving thing and a living thing.

_____ thing is a thing that grows, changes and needs food, air, and water to survive.

A _____

Memory Maker: Draw at least three different mammals.

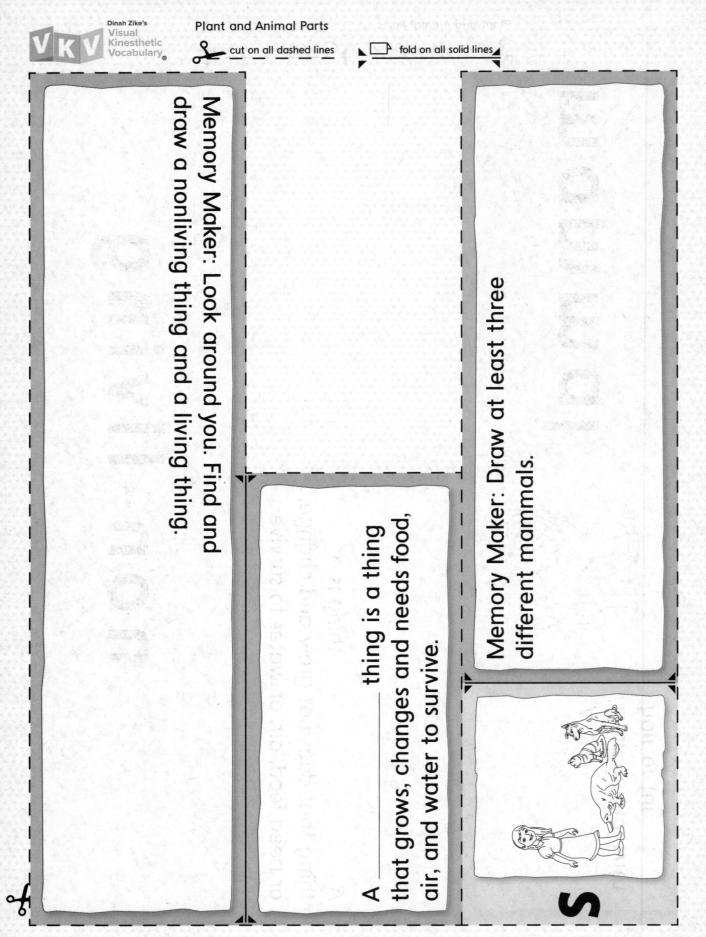

s

Copyright © McGraw-Hill Education.

Dinah Zike's
VKV
Visual
Kinesthetic
Vocabulary®

Plant and Animal Parts

✂ cut on all dashed lines

▢ fold on all solid lines

leaf

A _____ is
the plant part that
uses sunlight and
air to make food.

A _____ is
the plant part that
uses sunlight and
air to make food.

Plants use their
_____ to make
food.

A _____
is a part of
a plant that
makes seeds.

A _____ is
the plant
part that uses
sunlight and air
to make food.

The _____
is the part of
a plant that
holds up the
plant.

The _____
is a plant part
that keeps the
plant in the
ground.

Copyright © McGraw-Hill Education.

Memory Maker: Create, or draw, a plant that includes all plant parts described on this VKV.

flower

leaf

stem

root

Memory Maker: Draw plants you have seen that have leaves. Circle the leaves.

ves

Copyright © McGraw-Hill Education.

Dinah Zike's
Visual
Kinesthetic
Vocabulary®

VKV

omnivore

A **carnivore** is an animal that eats other animals.

1. A **herbivore** is an animal that eats plants.

2. An **omnivore** is an animal that eats other animals and plants.

Copyright © McGraw-Hill Education.

Memory Maker: Draw or write what each type of animal eats.

carnivore

omnivore

herbivore

herbi

carni

Copyright © McGraw-Hill Education.

Dinah Zike's
Visual
Kinesthetic
Vocabulary®

✂ cut on all dashed lines   ▭ fold on all solid lines

fall season

1. **Spring** is the season after winter.

2. **Fall** is the season after summer.

_____ is the season after winter.

_____ is the season after summer.

Copyright © McGraw-Hill Education.

VKV

Dinah Zike's
Visual
Kinesthetic
Vocabulary®

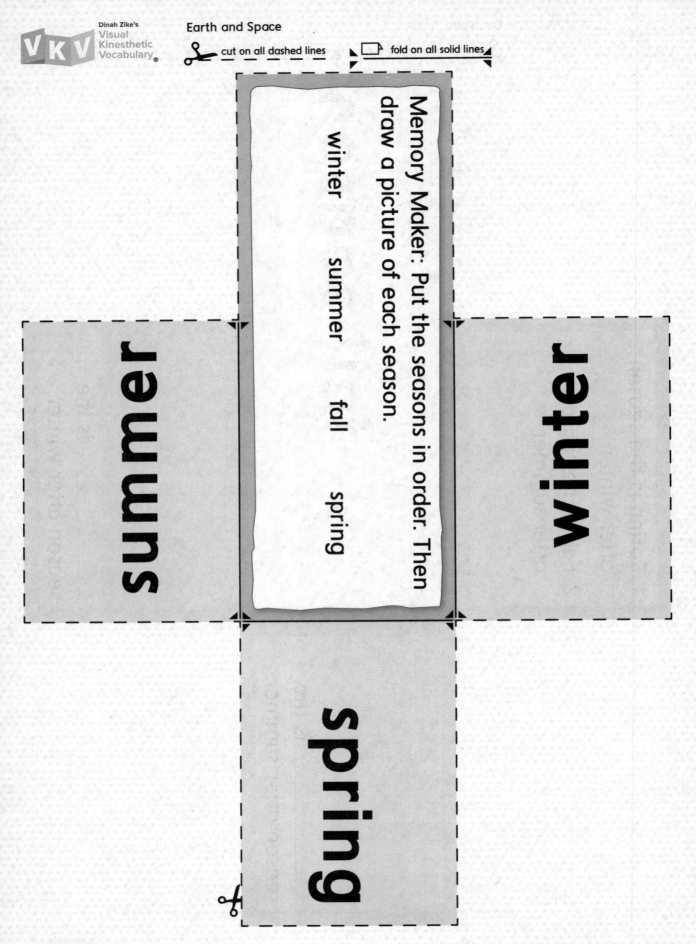

Memory Maker: Put the seasons in order. Then draw a picture of each season.

winter   summer   fall   spring

summer

winter

spring

Copyright © McGraw-Hill Education.

Dinah Zike's
Visual
Kinesthetic
Vocabulary®

Moon phases

_____ are the different Moon shapes we see each month.

day time

1. **Nighttime** is the time between when the Sun sets and rises.

2. **Daytime** is the time between when the Sun rises and sets.

Copyright © McGraw-Hill Education.

Dinah Zike's
VKV
Visual
Kinesthetic
Vocabulary®

Earth and Space

✂ cut on all dashed lines   ◄ ▢ fold on all solid lines ◄

Memory Maker: Make a story showing what you do during the daytime and nighttime.

night

The _____ is a ball of rock that moves around Earth.

Memory Maker: Construct a word web. Include all words on this VKV and any other words that relate.

Copyright © McGraw-Hill Education.

VKV Dinah Zike's Visual Kinesthetic Vocabulary®

✂ cut on all dashed lines

▭ fold on all solid lines

planet

A _____ is
a very large object
that moves around
the Sun.

A _____ is an object in the
sky that makes its own light.

The _____ is the star closest
to Earth.

The _____ can be seen
during the day.

Copyright © McGraw-Hill Education.

Dinah Zike's
Visual
Kinesthetic
Vocabulary®

Earth and Space

✂ cut on all dashed lines

□◄ fold on all solid lines

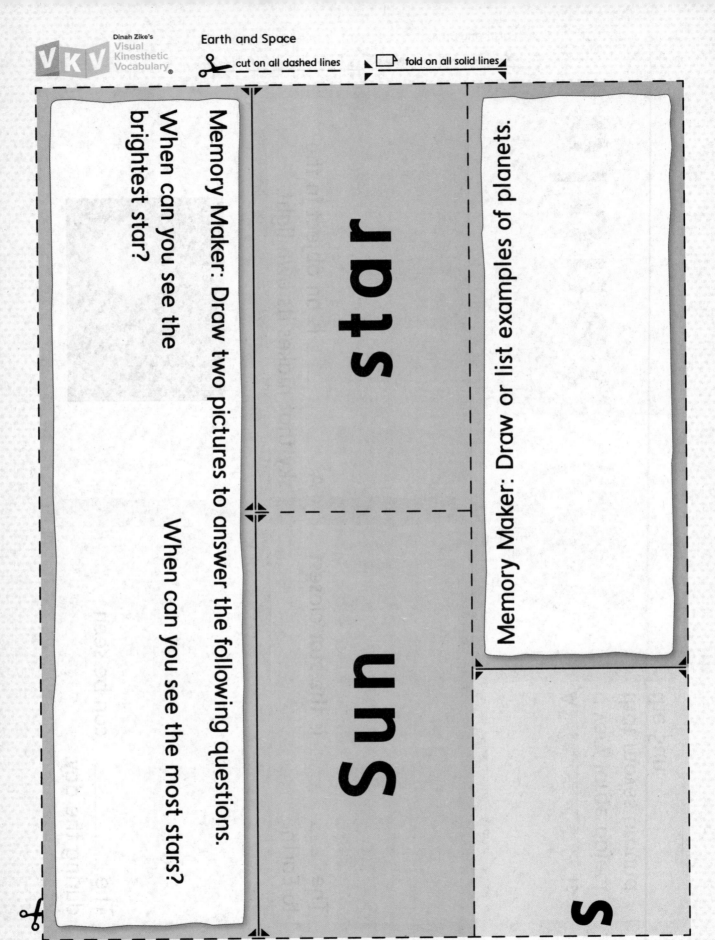

Memory Maker: Draw two pictures to answer the following questions.

When can you see the brightest star?

When can you see the most stars?

star

Sun

Memory Maker: Draw or list examples of planets.

s

Copyright © McGraw-Hill Education.